Science Fair Spelled W-I-N
Second Edition

A guide for parents, teachers, and students

by

Carl Tant

Biotech Publishing
A Division of
Plant Something Different, Inc.
Angleton, TX USA

Science Fair Spelled W-I-N
Second Edition

By Carl Tant

Published by Biotech Publishing
P.O. Box 1032
Angleton, TX 77516-1032

Printing 10 9 8 7 6 5 4 3 2 1

Library of Congress Catalog Number: 95 - 75899
ISBN   1-880319-12-8

Publisher's Cataloging in Publication Data

Tant, Carl
Science Fair Spelled W-I-N, Second Edition
YA 507.8
Bibliography, Glossary, Index, Tables, Charts
1. Science experiments
2. Biology experiments
3. Science Fair
4. Background information for parents & teachers
5. Student Level: Grades 7-12

Cover by Tammy K. Crask

Table Of Contents

Section VI: *Extended, "Add To" Projects: Next Year's Big Winners*

Section Section VII: *Quickies For Those Who Waited Too Long*

               1. Chromosome Staining
               2. Biuret Test for Protein
               3. Ninhydrin Test for Amino Acids
               4. Iodine Test for Starch

Glossary

References

Warning--Disclaimer

This book is designed to provide information in regard to its subject. It is sold for the use of middle and high school students with adult assistance and supervision. The background information is not by any means all that is known about the various subjects. However, references are provided for ease in obtaining further and more technical information.

The safety warnings and notes included cover only the most common areas of concern. There is no way to anticipate safety problems which might arise in a particular setting. Students of the target age group should never perform any type of scientific experiment without adult supervision. As pointed out in the first section, a good approach is to list all possible safety considerations which might occur with each experiment. As the student gets into doing the work, additional considerations should be added to the list if appropriate.

Because of the variables inherent in different environments and conditions for conducting the experiments, their success can in no way be guaranteed. The experiments are suggested as guides and should be adapted to fit individual situations.

As in common with any book, there may be mistakes, both content and typographical.

The author and Biotech Publishing shall have neither liability nor responsibility to any person or entity for any loss or damage caused or alleged to be caused directly or indirectly by following the suggested experiments in this book.

This book is sold with the understanding that the buyer will be bound by the above. If you cannot accept these terms of sale the book may be returned to the publisher in saleable condition for full refund.

Acknowledgement

A book like this is not written without the help and input of many people. Some who contributed above and beyond the call deserve a special thanks.

A few of my students at Angleton Senior High School who had survived the rigors of all the science courses offered earlier to reach the senior honors levels of Laboratory Management, Biology II, and Chemistry II assisted with reading and checking manuscripts. These include Glen Olinger, Leah Stern, and Wendy Turner. They were reinforced by underclassmen Brandon Marsh and Amy Van Wyk.

A very special expression of appreciation is due one of THOSE English teachers, Suzanne Muecke, who wields her red pen and caustic comments from the English Chair at Angleton Senior High School. She used at least several gallons of her red ink supply in reviewing the initial rough drafts of the manuscripts for this book.

Joe Vitale, publishing consultant offered valuable advice and counsel.

Monica Steves and Elsie Zikuhr of the Brazoria County, Texas, Mosquito Control District provided more information than we knew existed about mosquitoes as background for the illustrations in Section II Experiment 3.

While unnamed here, the many other students, teachers, and friends who offered suggestions, criticism, and encouragement will always be remembered.

Dedication

This book could only be dedicated to the one who made it possible initially by convincing me to undertake writing it when I thought there was no time. The world was suddenly reorganized to make the time. Her contagious zeal for the project, hard work, encouragement, cajoling, and occasional nagging have brought it to fruition. With inexpressible affection and appreciation this book is dedicated to my inimitable secretary, assistant, and I hope - friend - Renee' Walker.

Preface To The First Edition

A proliferation of new government regulations and other events in
recent years has made it more and more difficult to teach science and do
science activities. Rapid advances in new biotechnology have made it
difficult for both students and teachers to stay abreast of current events.
As new fields develop seemingly overnight, new vocabularies emerge.
Textbooks rapidly become obsolete before the end of their adoption.
Recently, pressures exerted by animal rights activists threatened to
eliminate the use of vertebrate animals in teaching biology. Rightfully
gone are the days when vertebrate animals could be used as experiments
of questionable value. Along with this, however, also gone, are such
simple experiments as asking classmates to taste two or three kinds of
cookies and rank them according to their preference. All of this, along
with MSDS for every chemical used, has turned off both students and
teachers from biology and chemistry science fair projects.

This book is about how to do it without such hassle. It is based on the
author's almost quarter century of science fair work as judge, sponsor,
and teacher. This is a book about intriguing valuable biology and
biotechnical projects. It is not a cookbook; it will require students to
think. It is about new technology on the cutting edge of the racing
advance of biological science. In addition, this is an honest--and
sometimes blunt--assessment of both students and judges. It is a book
about being confident and gaining self-satisfaction. It is a book about
WINNING.

The author and publisher wish you and our young scientists the best of
luck and hope to see you at the next International Science Fair.

Preface To The Second Edition

The gratifying wide acceptance and popularity of the first edition of Science Fair Spelled W-I-N indicates that it met a need. As its second printing neared exhaustion a decision was made to expand the book into a second edition with additional experiments and new references and to incorporate recent changes in science fair rules relating to group projects and home school participation. One request heard from many readers was to enlarge the section on report writing. This was considered, but only a few changes were made as the recent publication of a style manual for beginners (Williams, 1995) addresses that area more completely than could be done here.

Win more! Win again!

Publisher's Note

The use of the pronoun ,"he", with reference to science fair judges is not meant to imply that judges are only of the male gender. It is done only to simplify wording for students and avoid awkward sentence structure to include both sexes. The author and publisher recognize that most science fairs simply could not function without the highly competent contributions of female officers and judges. Well over fifty per cent of our capital stock is owned by women who concur that the approach used makes sense.

Don't get
Disqualified
by doing something
Dumb:

Some things are so common, we do not think of them with respect to their technical classifications. Most science fairs will classify the following technically, so get approval <u>before</u> starting your research!

Animal Parts & Products (Even if you pick them up off the ground.)

Eggs	Skin	Muscle (Meat from the grocery store)
Feathers	Saliva	Hair
Urine	Bones	Fingernails
Feces	Blood	Toenails

Chemicals

Detergents	Soap	Cosmetics
Waxes & polishes		Rubbing alcohol
Cooking fats & oils		Food coloring

Microorganisms

Bread molds   Fruit molds

Mildew   Yeast, cake and dried

Human Experimentation

Any physical activity.

Any mental activity. Yes, this includes survey projects such as, "Which of these three cookies do you like best?" or "Which of these four colors do you think is prettiest?"

Some of this seems beyond the realm of common sense, but there is nothing you or even the science fair officials can do about it except abide by the rules. You have probably concluded by now that some of the more ridiculous items originated in the minds of various bureaucrats. No further comment.

Section 1

Hints and Helps:
Elation, not Frustrations

Introduction

This section contains basic general background information about science fairs. Do not feel that you should not read the chapters directed to someone other than yourself. Students, teachers, and parents can all find some valuable information in all parts of this section.

You will find much additional useful information at the beginning of each section. The title pages for the sections list the science fair classifications which would be most suitable for the various projects. Where more than one is listed, your choice will depend upon the emphasis your individual project utilizes. You will also find a listing of related areas of knowledge. Any special safety precautions, in addition to those normally observed in good laboratory practice, are given for each experiment.

Following are the most common science fair classification abbreviations.

BSS	Behavioral/Social Science
BIO	Biochemistry/Microbiology
BOT	Botany
CHE	Chemistry
ENG	Engineering
ENV	Environmental Science
ESS	Earth/Space Science
MED	Medicine/Health Science
MTH	Mathematics/Computer Science
PHY	Physics
ZOO	Zoology

Some large science fairs may split the combination categories such as BIO into separate groups. Be sure to check the current ISEF and your regional fair rules.

Chapter 1

It's More Fun To Win

Why are you reading this?

Perhaps it is because your teacher just aimed a loaded cannon at your head and announced, "Everybody is going to do a science fair project." Perhaps you have found a subject you are really interested in and want to learn more about. Perhaps you have an inborn urge to do scientific research. Perhaps you looked at the long list of scholarships and other awards and decided you wanted some. Or, perhaps it is a combination of many of these.

If you are just doing it because you are being forced to and really don't care about it, then find a simple project, throw together the minimum acceptable to keep your teacher off your back, and be prepared to live with yourself. If you are going to do it for other reasons, think about what has been said about winning: "It is not so much if you win as it is what you learn." That statement is a wonderful idealistic truism. No one can argue with it, but let's get real. It's a lot more fun if you do win. And that is what this book is all about.

Let us first consider some things that really aren't science--except maybe the science of people.

The Judge

Many students view the judge with horrible fear, trembling, butterflies in the stomach, sweat on the brow, etc. Don't. Judges are human--you know, *Homo sapiens*, like the rest of us. In age they range from over 80 down to last year's science fair grand award winner.

Most judges are there because they want to be. They are interested in young people and interested in tomorrow's scientists and engineers. A few are there primarily because their bosses told them something like, "Our company needs to supply a judge for the science fair. I'm glad you are volunteering." You might find these involuntary judges a little grumpy, but you will find most of them fair and honest in their evaluation of the projects.

Another category includes the one whose alarm clock did not go off that morning and, rushing to work, received a speeding ticket on the way. When this particular judge left work that afternoon to go to the science fair, he or she was further delayed by a flat tire. Do not expect this judge to like your project. Be consoled by the fact that this judge will not like anyone else's either. Probably the highest score anybody will receive will be a 30.

The worst--or the best--judge you could get would be the one whose career specialty is exactly the subject of your project. He will either spend 30 minutes telling you everything you did wrong or giving you many ideas about improvements you could make for the next year. Give all of this your careful attention. Use whatever is appropriate to snow the next judge with your vast knowledge.

Read the judge. Follow the judge's lead with appropriate action. Judges differ tremendously in evaluating a project. Some will approach you, grunt to acknowledge your presence, and then stand without saying a word until they have read everything on your project display. Others may glance at your posters, ask for a written report, read every word of it, and finally start talking. Many students would classify all of these as varieties of the genus and species *Mynervus reckus*. Another type looks at your poster and after a few seconds starts pointing to various sections of it, rapidly asking questions such as, "What do you mean by this? Where did you get this idea? How do you know this data is correct? Are you sure this is the best way? What makes you think this is really the answer?"

The other kind of judge is the one everybody likes. He comes up with a smile, introduces himself while glancing at your display, says something like, "Tell me about your project; I'd like to see what you've done." For these judges, have in mind a brief, pertinent, very much to the point summary of your project. Plan to do it in about two minutes. That will give the judge ample time for questions or pursuing parts that he finds particularly interesting.

It is a good sign when a judge spends more than the usual time with your project. He may ask a hundred questions, but remember that while he is with you he is not looking at someone else's project. He has only a limited amount of time. If your project keeps him tied up, somebody else's may get short shrift. Dirty pool? No, it is simply reading the judge, following his lead, and playing to win.

Science fair officials spend much time and effort recruiting fair, honest, competent, and interested judges. But, again think reality. A large science fair may have seven or eight hundred different judges. Think about the number and kinds of personalities in your class of 25 or so students. Multiply that number by 28 and you will have the number of different kinds of science fair judges you might meet. Be prepared to deal with them accordingly. Remember, they are dealing with perhaps an equal number of different student personalities. This is part of what makes life so interesting.

"I Don't Know"

The dreaded moment is going to come. The judge is going to ask a question you can't answer. If for no other reason, judges will keep asking questions to test the limit of your knowledge. When the time arrives, do not be afraid to say "I don't know."

Consider this a golden opportunity. If the judge does not volunteer the answer, ask what it is. Store this in your memory bank to use to impress another judge. The first judge will be flattered and impressed with your desire to learn more. After all, that is supposed to be one of the purposes of science fair.

Many times a judge will ask a question to which he does not know the answer. The racing advances of biological science have made it highly specialized in the last few years. It is impossible for one person to keep up with the cutting edge of each new technology. The judge may be asking a highly technical question simply because he is curious and hopes to learn something himself. If you have aroused his curiosity, the judge may go off and discuss it with others who might be more specialized in your area. He will probably return later to give you the information.

Practicalities of Life
at Science Fairs

Here are a few other ideas the author has picked up in over 20 years of working with science fair participants.

First impressions are crucial. Judges usually are working under strict time limitation. Their initial impression of you and your project will result in at least a subconscious favorable or unfavorable attitude. Be ready and be on your feet with a pleasant, but not faked expression when the judge approaches. Again, follow the judge's lead in presentation.

Part of the first impression is your appearance. If you have not been to science fair before, ask your teacher about appropriate mode of dress in your location. Sloppy casual dress is on the no-no list. Dress tastefully to be pleasing to many different kinds of people. Above all else, avoid high heels or other uncomfortable shoes. You might be on your feet for ten to twelve hours during the judging process. Another problem to prepare for ahead of time is temperature variations. The heat or air conditioning may be too high, too low, or whatever. It is a good idea to have something extra to put on, or to wear a jacket or blazer that can be removed.

"Erk, Glurk, Glumps, Blup." Surely you can recognize that as the project explanation made by a student who had filled his mouth with sticky caramel candy just before the judge came up. A judge has not been seen in your aisle for at least half an hour, so you decide to ease the hunger pains a little. This is guaranteed to make a judge materialize suddenly from nowhere. If you must nibble, at least do it with something that you can swallow quickly as the judge comes up. Your project table is no place for a soft drink can. It could easily get knocked over and mess up months of work. It only takes once. There will be many moments or, perhaps, even a few hours of boredom when judges are not coming by, but be very careful about cluttering up the aisle with board games or other entertainment shared with your fellow contestants. A judge will not have a good impression if he is forced to lean over games in order to look at your project.

Leave the music outside. Even if your science fair does not prohibit bringing portable radios, cassette players, etc., you should have the good sense not to do so. Again, first impressions are vital. That few seconds spent removing earphones or turning down the music can make a terrible start with the judge.

"I Gave A Party, But Nobody Came"

That header might well be the feeling of the judge who comes to your project but doesn't find you. He might not return. He could simply mark the scoring sheet, "student missing" and go on to another project. There will be plenty of time to meet new friends, socialize, or view other projects when judging is not in progress.

So what if you absolutely, have to, cannot wait to go to the restroom? Try to pick a moment when judges are not close by and ask the contestant next to you to tell any judge that shows up that you will be right back. Then be right back. Further hint: quit guzzling so much carbonation, and this absolutely must go time will come less often.

A Final Thought

Above all, do not be upset if every judge does not enthusiastically fall in love with your project. You cannot expect to please everyone. A science fair is a numbers game. Consider the overall big picture. There will be other judges who do not like other projects. It all balances out, and that is the purpose the directors have in mind when they set up the judging with many judges. Fairness must take into account human foibles.

After you have been out of school for a few years, you will look back and know that one of the most valuable things you learned at science fair is how to deal with a wide variety of people on a moment's notice and with no previous acquaintance. This will stand you in good stead with your future employment as well as further education. In addition to personal impressions, another major aspect of winning is your project display. We get into that in the next chapter.

Chapter 2

The Project Display

You have probably been told that the project display itself counts only five to ten per cent in the judging. That may well be true, but do not place too much stock in it. Again, judges are humans. The initial subconscious impression is vital in forming the attitude you want. Here is another chance to manipulate the judge's mind in your favor. Does that sound terrible? Perhaps so, but it is a tough competitive world out there. If you are going to play the game, by all means play by the rules, but also play to win.

Figures 1 and 2 on the next page show projects which were recent major winners in the huge highly competitive Houston Regional Science Fair. These displays speak loudly. Among other things they tell the judge is that your research was well thought out and well organized. These posters were obviously not thrown together the night before the science fair.

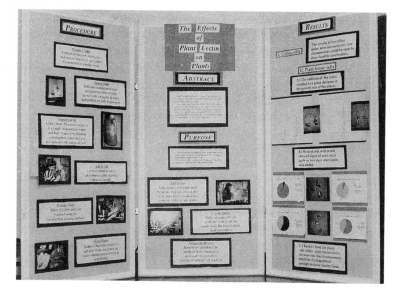

Figure 1

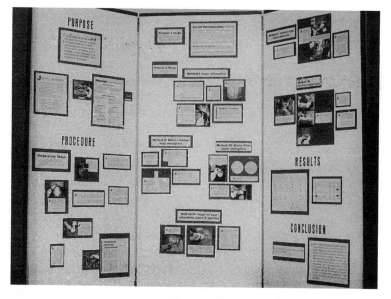

Figure 2

Some Practical Pointers

Your data may require only a two foot high display board. Size is not supposed to count in the judging, but let's face the facts. If a two foot display happens to be positioned between others that are four feet high, the smaller one will be dwarfed and look insignificant. On the other hand, you do not want a large board with a lot of empty space. That gives the impression you haven't done very much. This is where mounting tricks become vital as evidenced in Figure 2.

Do not overcrowd your display. Even if you have stacks and reams of data, avoid trying to put it all on the board. A jammed-up board will not get lengthy attention. Have your backup data easily available in neat binders or other arrangements to show the judge on request. Stick to a summary on the display itself.

Remember that, among other things, your display is your advertising. It must attract attention and say, **"LOOK AT THIS PROJECT-- IT IS GREAT!"**

One way to get good ideas about display arrangement is by studying ads in high quality magazines. Notice the use of color and arrangement to attract attention to the central theme.

Be cautious about color. Make certain that the colors you use go well together and look attractive. Avoid colors that are too dark and will not stand out well if your project happens to be placed in poorly lighted area. Likewise, avoid fluorescent poster board and markers. A gaudy, cheap carnival atmosphere about your project will not score points. Simplicity and good taste make a strong statement of overall quality.

Follow The Conventions

Judges will expect to see the standard steps of scientific research on your display. These include:

Problem. A brief statement of your subject or question.

Hypothesis. Your guess of how to answer the question asked.

Procedure. A short summary of the steps in your research.

Results. State and summarize your results.

Discussion or interpretation. This is the place to state how your results apply to question and hypothesis.

References. A brief summary of pertinent literature used or referred to in a long bibliography you have elsewhere.

Acknowledgement of assistance. Almost all scientists receive assistance and suggestions from others about their research. This is the place to list those who assisted substantially.

Possible Pitfalls

Select your materials carefully. Three two by six feet by three-fourth inches particle boards get very heavy and hard to handle. Think about some of the new light weight foam-core boards for your display. Avoid those which are less than three-eights inch thick. They will tend to bend and curl unless supported from the back. Arrange your boards so that they will come apart and go together easily or fold flat, while at the same time protecting mounted materials.

Protect your display while loading, unloading, and in transit. It takes only a few drops from a sudden rain shower to totally ruin an attractive display. Put your boards in a large trash bag or otherwise cover securely with waterproof plastic. If your display is large, you might want to think about wrapping it in a painter's polyethylene drop cloth as an inexpensive protection.

Many judges over the age of forty come equipped with bifocals. (You will, too, someday.) This means do not put small lettering or drawings at the top of a tall board where they will be difficult to see. Likewise, be careful about positioning other small materials in such a way that they are difficult to see. A judge will not look long at the cost of a crick in the neck or a slipped lumbar disk. Any standard typewriter size print on the display is of questionable value.

Keep it simple. Put the mass of detail and data in notebooks or presentation folders on the table. It is a good idea to have separate ones for different topics. This also offers you an opportunity to pick up the material and place it in the judge's hand so that he will have to look at it. That is a salesmanship trick, but it will keep the judge with your project longer.

Another important consideration is the materials that go on your display board. Some glues do not stick well to some surfaces. Try them out for a day or two before you make the final application. Make sure that the glue does not bleed through and cause spotting of various colored paper or inks. Also, check to make sure that it does not cause your materials to wrinkle.

The rules say that your project display must be self supporting. Make sure that it is, particular if you use one of the new light weight foam boards. It might be worthwhile to put a piece of tape behind it to help hold it to the table top or use some other support or base such as a short length or two by four lumber. If your display falls over on a judges head, you might as well forget it and go to a movie or go home and think about next year's project.

The Murphy Box

The author's students long ago learned that Murphy's Law always functions at science fair. "If anything can go wrong it will." Therefore, be prepared. Carry along an assortment of items needed for last second repair or touch-up. Otherwise, these things may not be quickly available when you need them. Include scissors, two or three kinds of tape, an assortment of felt tip markers, pens, pencils, glue, thumbtacks, and some basic tools such as hammer, screwdriver, and perhaps even a small handsaw. If your display uses electricity, carry along spare lights and an extra extension cord.

Take a small notepad. It will be handy for writing down information obtained from judges or fellow students. It also might be very useful in writing names, addresses, and phone numbers of interesting students of the opposite sex that you meet during the fair!

By the way, if one of the fringe benefits you expect to get from science fair is filling up your address book for potential new dates, make sure to jot the subject of the person's project down by the name. If you collect very many, after a few days it is easy to confuse who did what. You do not want to call up this attractive person and start the conversation by talking about someone else's project. Your score for that will be minus ten.

Stand Alone Presentations

In some science fairs, including the International, there is a period of time when judges evaluate the projects without the student present. If your fair has such judging, be sure that all your data and anything else you want the judge to see is neatly arranged on the table and very clearly labeled. This type of judging needs a very detailed report properly written in a scientific manner.

Along with the physical display of your project, the judge will also consider the quality and clarity of your oral presentation, if he asks you to describe what you have done. If he simply uses the approach of questions, he will consider coherence, quality, and succinctness of your answers.

Project Objectives and Design

The judge will not deduct points if your project is the same subject as others in the fair or if it repeats work previously done by others. However, you will score much higher if you have an unusual and creative objective expressed clearly. Originality of approach--in other words, modifying or finding a new way of doing something--counts heavily and will score extra points for you.

Continuing Projects

Projects extended and continued from previous years should be clearly labeled as such. The judge is required to ascertain the portions which are the current year's work and base scoring solely on it. He will, of course, note the foundation laid by the earlier work and consider it as a type of reference or basis for the current research. He will certainly look very closely at your proposals for further extension next year. If your planning has progressed and you have enough data to justify it, including a protocol outline for future work would be of value.

Team Projects

The international science fair has recently added a new category of "Team Projects" which includes all categories. Team projects compete with each other. Here are the regulations, reprinted with permission, from the 1995 International Science & Engineering Fair Rules.

"Teams may have up to three members. <u>NOTE</u>: Teams may not have more than three members at a local fair and then eliminate members to qualify for the ISEF. A team project cannot be converted to an individual project. A new member cannot be added to a continuing Team Project, but two original team members may continue their research if the third member no longer participates.

Each team should appoint a team leader to coordinate the work and act as spokesperson. However, each member of the team should be able to serve as spokesperson, be fully involved with the project, and be familiar with all aspects of the project. The final work should reflect the coordinated efforts of all team members and will be evaluated using the same rules and similar judging criteria as the other 17 categories. The team members must jointly submit one abstract and one research

plan that outlines each person's tasks. <u>Names of all team members</u>
<u>must appear on the abstract and forms.</u>"

Home School Participation

Students who participate in home schooling should check with their local or regional fair director for information about participating. If you do not know the nearest fair location, you can obtain it from Karen Royden, Director of Youth Programs, Science Service, 1719 N. Street NW, Washington , DC 20036

Chapter 3

What The Judge Is Looking For

Your attractive project display, charming personality, and remarkable personal good looks have attracted the judge to your table. Now what?

Scientific Thought

All judges, regardless of the way they were classified in Chapter 1, will be looking for certain things in common. At the top of the list, counting perhaps as much as 30 per cent in the scoring, will be **Scientific Thought and Procedure**. The judge will be looking for evidence of a properly organized approach to solving your problem. This includes a logical progression of work from one point to the next. Having an initial protocol helps here. The protocol is nothing more than an outline of what you plan to do. The judge will recognize that in many cases it is necessary to modify the proposed protocol as evidence begins to accumulate. Do not be afraid to show these modifications. They are proper and expected.

This is one of the places where the judge will look for evidence of a review of scientific literature related to your subject. Your bibliography is important. The more references you list, the better off you will be. The judge will be impressed by occasional citations of specific articles in your report. If you have one or two outstanding references, you might want to cite them in your project display. Further evidence of scientific thought should be in the discussion section of your report. It is good to have a paragraph describing various alternative approaches you might have used. Be sure to state clearly why you chose the one used. If the judge is not one who reads reports, be sure to point this out in your oral presentation.

Another item which demonstrates scientific thought is a short section which might be entitled "Future Work." In this you should point out possible extensions of your work based on the results you obtained and reported here. This is particularly important if your project is one you plan to expand and continue over the next two or three years. No one will expect you to follow these projections exactly. It is recognized that changes will be necessary as you go along. However, showing this item is strong evidence of thinking ahead.

This is also where the judge will consider your discussion and interpretation of results. Be especially careful not to draw conclusions which are unwarranted by the evidence you have obtained. Above all, do not base conclusions on insufficient evidence. Often it is necessary to refer to such data. If this is the case, use hedge words such as "possibly indicates" or "might be..." or "with more testing might lead to ..." The judge will properly interpret such statements as indicators of your awareness of the need for more evidence to support firm conclusions.

Keep your project penned up in its original corral. It is all too easy as your research progresses to go chasing off after new areas that become possibilities. Do not yield to this temptation. These areas are proper subjects for future work. If you spread yourself too thin and attempt to go in too many different directions at once, your score will be lower.

Thoroughness

The next big bite on the judges score card is in the area of thoroughness. This is the aspect of science fair projects most often and most severely criticized by judges. It is far better to cover a very narrow subject area totally and completely than to spread a wide assortment of data into related areas.

Thoroughness means significant numbers. That means repetition. Do not conduct any aspect of your experiment only one time and attempt to draw conclusions. The minimum is at least two or three repetitions. If the results are identical, you have an arguing point.

Thoroughness also means trying the same thing in different ways if such is possible. For example, if you are analyzing the protein content of a substance, do your analysis by at least two different methods. Sometimes there will be a difference in results depending upon the method used. Be sure to make this aspect of your work obvious to the judge.

Goal Achievement

While it does not count many points in the scoring, the judge will be looking for achievement of your stated goal.

This one can be easy. If you cannot reasonably predict the result of your experiment, it is wise to state your subject or hypothesis in the form of a question which can legitimately and honestly be answered either "yes" or "no." That way you will not have to contend with the problems of negative evidence or admitting your hypothesis was incorrect. Either the positive or negative answer will be a proper outcome.

Related Knowledge

Learn all you possibly can about any area even remotely related to your primary subject. **Judges will be extremely impressed if you can refer to pertinent literature and subject matter by author, title, and date**. If necessary, memorize a few of these. It is particularly important if some of the sources you refer to are those of persons considered to be important or outstanding in the field of your subject. Go back in time. Judges like evidence of historical knowledge of your subject, particularly if it points to improvements or changes made in early theories. A good example of such presentation might include statements like, "In 1901, Smith did so and so. In 1910, by doing such and such, Jones confirmed Smith's earlier theory."

Competition can be tough. There probably will be many good projects in your fair. The scores separating first place and honorable mention may differ by very few points. A thorough knowledge of the scientific literature related to your subject can tip the balance in your favor. *Read, read, read, read--remember and quote!*

Project Display

This one has been thoroughly discussed in Chapter 2. While it usually counts ten per cent or less in the overall scoring, it is vitally important from the standpoint of psychologically influencing the judges.

Along with the physical display of your project the judge will also consider the quality and clarity of your oral presentation if he asks you

to describe what you have done. If he simply uses the approach of questions, he will consider coherence, quality, and succinctness of your answers.

Chapter 4

Writing The Report

A formal scientific research report should be an integral part of every project. Unfortunately, the report is often a major source of student stress. The first question asked is usually, "How long does it have to be?" This is immediately followed by, "How do I do it?"

The English department or the school library should have several good books about scientific writing. Some examples are given in the references. One of the best for the inexperienced is the recent book by David Williams (1995), **Writing Science Research Reports - An Introductory Step-by-Step Approach to A's**

Along with these general guides, the customary format for any science specialty should be followed. Most scientific journals have instructions to the author in each issue or in the cumulative index. Some general guidelines to use as a starting point are shown on the following page.

All reports should be typed, double spaced, and grammatically correct according to current formal scientific usage. If the report is written on a word processor, be sure that it is printed with easily read type. Avoid the draft quality nine-pin dot matrix done with a worn out ribbon. Most judges have no complaint about good NLQ done with a dot matrix printer. Reference note cards or appropriate computerized reference forms should be available in a separate folder or binder for the judges to check.

Report Organization And Format

The most universally used format includes:

1. **Title page**
2. **Abstract**
3. **Introduction**
4. **Literature review (may be combined with introduction)**
5. **Experimental procedures (materials and methods)**
6. **Results**
7. **Discussion of results**
8. **Future work**
9. **Acknowledgement of assistance**
10. **References**

Judges do not expect science students to be expert typists. They do, however, have a right to expect an absolute minimum of typos and no misspelled words. Spelling will probably have to be checked the old-fashioned way. Unless a spell-checking program is designed for a specific field, it is virtually useless and will stop several times in each paragraph with the message "I don't know that word."

Enlist the cooperation of an English teacher to review a final draft. This will be of great value in helping spot dangling participles, run-ons, split infinitives, and other no-no's. The English teacher will appreciate being provided a copy of the writing style guide for the specific scientific discipline involved.

Writing Style

Scientific literature reviews and reports of original research are ordinarily written in past tense, third person, passive voice. It is permissible to change to present tense where appropriate in the "discussion of results" and "future work" sections.

Under no circumstances are personal pronouns such as "I," "we," or "our" acceptable. For example, say "results of such and such were obtained," instead of "we obtained such and such results." The use of first person cannot be defended by saying that it appeared in some science journal. A competent judge's reply to that is likely to be something along the order of "so the editor got as sloppy as the writer." Incidentally, most English teachers do not like third person, past tense, passive voice. Journalism teachers have heart attacks when they see it. Remember however, in Rome...

Exercise caution in the use of abbreviations. Only those listed on page 180 or so designated in other tables are considered standard scientific abbreviations which may be used without definition, as it is assumed that all competent scientists are familiar with them. Non-standard abbreviations should be defined with their first use. Such abbreviations are a hindrance to scientists specialized in fields other than that of the author. Their use should be held to a minimum.

Maximum use should be made of standard chemical symbols with numeric subscripts for naming chemical compounds. Short abbreviations are almost universally used for some biochemical compounds such as ATP, NAD, RNA, DNA, RNAase, DNAase, etc. You will find the ones most commonly listed in the appendix "Abbreviations" in this book.

Macromolecules of repeating sequences may be represented by the prefix "poly-," indicating that it is a "polymer of." Particularly with reference to carbohydrates, "oligo-" may replace "poly" in some names.

Abstract

LECTIN ISOLATION METHODS

Scientific studies lack some vital information about lectins--carbohydrate binding, cell agglutinating proteins. Relatively few lectins are known and research is lacking to determine which methods of isolation and concentration are best.

In this research, a lectin was isolated from the plant Xanthosoma mafaffa, Six methods of isolation and concentration were tried. For this lectin, the most effective methods involved hog gastric mucin adsorption.

Lectin characteristics were confirmed by red blood cell agglutination with three species of mammalian blood.

Pertinent literature related to characteristics and biochemical behavior of lectins is reviewed.

Fig. 3.1

Abstract

The abstract should be a brief informative summary of the content and conclusions of the paper. It should specifically mention new methodology and chemical compounds. Between 200 and 300 words is a good length. Procedural details should not be included. The abstract is intended to give the reader quick information about the purpose and significance of the research. It should be intelligible to the non-specialist in the field and hence must avoid specialized terms or abbreviations that require definition. It should not contain information which is not supported in the body of the paper. Figure 3-1 is an abstract from a project at a recent Houston Regional Science Fair.

Introduction

The introduction should briefly state the purpose of the investigation in its relation to other work in the same field.

If only a few literature references are available, they may be incorporated in the introduction. If an extensive literature review is required, it should be presented separately, and only a few references of a general nature should be included as part of the introduction. A transition statement is usually not required to conclude the introduction in order to move into the main body of the text.

Literature Review

If extensive literature is available, it should be reviewed and described in either chronological or subject order. The writer should not include personal comment or interpretations at this point. References to specific procedures may be included in a general way, but should be cited again and described in more detail where applicable in the methodology section.

References for most biology research should be cited in the text by author followed by Arabic numeral in parenthesis or author followed by date in parenthesis. In some cases it is permissible to give the date in the form of a personal phrase in commenting on a work. Text citations to references written by more than two authors should be styled as "Smith, *et al.*"

The use of "in preparation," "private communication," and "submitted for publication" is not allowed in the reference list. "Personal communication" or "unpublished work" should appear parenthetically following the author's name in the text. A "personal communication" should only be referred to upon written permission of the person who has made it. Since the reference to a paper "in press" implies that the paper has been accepted for publication, the name of the journal in which it will be published must be given.

Abstracts of papers presented at scientific meetings should not be cited in the list of references unless they have been published. Abstracts not citable in this manner should be included in footnotes.

If the report cited is one of a series, it should be identified as such in both the text and the list of references.

If the review of the literature includes only a relatively few papers directly concerned with methodology or data interpretation, these may be included at appropriate points in the text without the necessity for a separate literature review section. Each case should be judged individually on this point.

Experimental Procedures
(Materials and Methods)

This section should be brief as possible but must provide adequate information for repetition of the work by a qualified person. Previously published procedures and their modifications should be cited. If a well-known established procedure is used, reference should be made to common standards.

The author should draw attention to any potential unusual biological or chemical hazards which might exist. If the work involves any government regulated materials or methods, reference should be made to the appropriate NIH guidelines, OSHA, EPA, Dept. of Agriculture regulations, etc.

If modifications of procedures are made during the course of the work, the modification as well as the original procedure should be given.

If relatively few different types of materials were used, they may be listed and identified at appropriate places in the procedure description. In this case, the manufacturer or source of the material may be given in a footnote; otherwise, descriptions and sources may be given as a separate sub-section under procedures. Tradenames and trademarks must be cited. In biological work, sources of cell and tissue lines must be given.

All chemical names should conform to the preferred IUPAC nomenclature rules. References to simple chemical compounds may be made in the text by formula only when it can be printed in a single horizontal line of type. Structural formulas and equations should be centered between successive lines of type.

Chemical concentrations should preferably be expressed as molarity. Molecular weight should not be expressed as daltons. Molecular weight is the relative molecular mass (M) defined as a mass ratio which is a pure number. Instead of "the molecular weight of X is ten Da," one should say, "the molecular weight of X is ten." It is correct to express the mass of an entity of undefinable molecular weight in terms of daltons, as "the 12,000-Da protein."

All culture media should be titled with the complete unabbreviated name, and unless it is a new medium designed by the author, the source should be cited. In either case, the complete composition must be given. If several media are used, particularly if they are variations of each other, the use of a combined comparative table might be appropriate for presentation.

The composition of all buffers should be defined to specify the concentration of each species and the pH at which it functions. For simple buffers such as 0.1 M sodium acetate (pH 5.0) it is assumed that the molarity refers to the total concentration at the specified pH. The composition of mixed buffers is shown by the use of a colon or solidus between the components and their respective concentrations. The use of hyphens is not acceptable. Do not use abbreviations for buffers. They may be designated as Buffer A, Buffer B, etc. after being defined as such.

All enzymes must conform to standards as specified in the latest edition of <u>Enzyme</u> <u>Nomenclature</u>.
All organisms must be specified by the proper Latin binomial name. Genus names should be abbreviated after first being used in full.

Results and Discussion

If the results are sufficiently brief and little discussion is required, these two items can be combined and presented together in a paragraph form. Such could be appropriate even if one or two tables of graphs are needed.

Lengthy descriptions of results or data which require tabular or graphic presentation should be given as a section separate from the discussion.

Discussion should be limited to interpretation of results and application of results to related areas or to previous research. The means of overcoming any difficulties encountered during the research should be described. Where appropriate, it is desirable to conclude the discussion with a brief statement of future work which might be done to expand, apply, substantiate, or clarify the present research.

When And What

Good reports take time. They must be written and rewritten. They are not written at 2:00 a.m. on the day science fair starts, nor are they written with a portable typewriter on the bus carrying students to the science fair. Start early.

Some judges may not pick up the report. If the report has been thoroughly done, the judge's attention should be called to it. This might be done simply be pointing it out, picking it up, and placing it in the judges hand. It can also be accomplished by referring to some data with a statement on the order of, "all this information is summarized here in my report" as the report is opened to the appropriate page and handed to the judge. Almost equally discomforting to the student is the judge who will stand and read every detail of the report without uttering a word for ten minutes. This is the time to simply relax because very little, if anything, can be done with that person. Be prepared for his questions about specific points in the report when he does start to talk.

References

References in the text should be cited as in the Literature Review Section. They should be listed at the end of the paper in numerical or alphabetic order. Journal titles should be abbreviated to conform to the system used by Chemical Abstract Service of the American Chemical Society. A few journals still follow the older convention of listing references in alphabetic order and date by the last name of the senior author. Adjust the bibliography format to that of a major journal in the field of the research.

References which do not contain the title of the paper cited are not acceptable.

Format the references according to the following examples:

For a journal article:

Kao, K.N., F. Constabel, M.R. Michayluk, and O.L. Gamborg, 1974. Plant protoplast fusion and growth of intergeneric hybrid cells. Planta 120:215-217.

An immediately following reference by the same senior author would have his name indicated by dashes preceding the other author. For example, if a paper by Kao, Smith, and Jones followed the above, it would be started:

-----, A. Smith, and B. Jones, 1985. ....

For a book:

Otto, J.H. and Albert Towle, 1985. Modern Biology, 9. 271. Holt, Rinehart, and Winston, New York.

For a chapter in an edited book:

Constabel, F. and J. A. Cutler, 1985. Protoplast Fusion, in "Plant Protoplasts" (Fowke, L.C., ed.) pp. 53-65. CRC Press, Inc., Boca Raton, Florida.

For published proceedings of a symposium or short course:

Treat as any of the above appropriate to the format of the publication.

While it is proper to use the *et al* designation in the text for papers with multiple authors, it is improper to do so in the references listing where full names or initials must be supplied.

Tips For Teachers

Your school administrators have just decided that participation in science fair would be a major step forward in the enhancement of science and math education. You have been selected as sponsor. You have already said, "I have never done that--I don't know anything about it." They turned deaf ears on your pleas, and now you find yourself wondering how to go about sponsoring science fair projects.

The task is a formidable one. It is also a rewarding challenge. You will share with your students and their parents the major problem of finding time. You will, along with the parents, agonize over the question "How far should I go in helping?" The time to get started approaches, and you wonder what to do first.

These are not questions with simple answers. Like many other aspects of teaching, they seem harder the first year. The first year really isn't harder--it is simply a means of learning how to handle science projects.

The basic guiding principle to keep in mind is that the project is the student's. It is the student who must face the judges. Your job is to lead, direct, and teach your student how to do it right.

Selecting A Subject

You will be a member of a fortunate minority if your student has already decided on a topic. For most, selecting the subject is a major hurdle to be overcome initially. Before the final topic is picked, the student may receive suggestions from many sources--friends, parents, acquaintances from business, etc. This is your golden opportunity to counsel the students that they are embarking upon what might be a time consuming experience that will last through high school, and may even serve as a springboard to the future. Emphasize the selection of a topic which is of interest to the student. Nothing will be gained from the effort and sacrifice of social time or other activities if the project does not really capture the student's imagination and interest.

There are literally thousands of sources of ideas. Contemporary news stories in papers and periodicals are an ongoing source. Most students are interested in being on the cutting edge of new technology as it develops. Any of the popular science periodicals are good sources of ideas. A step beyond those would be journals such as *Science* and *Scientific American*. One of the best sources for biology ideas is *Biology Digest*. This review journal is published for students monthly during the school year. It contains abstracts of current research reports from the scientific journals. These are written in a language understandable by high school students.

Another good source of ideas is the listing of projects presented at previous science fairs. A related starting point is the list of research submitted to the Westinghouse Science Talent Contest each year. Most laboratory manuals have experiments which can be extended into worthwhile research projects.

You will find it worthwhile to start a small notebook or card file of ideas as they occur to you. Sometimes they escape with equal ease. As your collection grows over a few years, you can become more sophisticated with a data base computer program for it. Making this available to the students will permit them to browse for ideas, select a few with possible interest, and discuss those in more detail with you.

Another source is the science specials presented from time to time on television. Many encompass topics which could be expanded into well-designed research projects. Finally, do not overlook good science fiction! Much of today's science in biology was considered science fiction in the 1970's. Good ideas can be developed from many of the situations described by competent science fiction authors.

Sooner or later will come the time to consider the merits of basic versus applied research. The question is simply unavoidable. Overall, most judges make every effort to be fair and objective. However, when several projects have close scores, the personal backgrounds and feelings of judges must become at least a subconscious factor in final decisions. Academicians will naturally gravitate toward basic research projects; judges from industry will have a background which involves applied research. This is a good time to look at the awards given in your science fair over the last two or three years. Categorize the winning projects as to basic or applied. Go over your results with your students and let them decide which way to go if the question is of concern.

Placement of the project in the proper competitive category in the science fair is another matter of great concern to teachers. Obviously, some projects could be entered in only a single place. A quick glance through the projects in this book will show that many others could legitimately be directed toward several areas. This subject should be considered months ahead of science fair, particularly if your school will have many entries and the number per category is limited. Emphasis on different areas will permit you to have more students. Do not stretch categories too far, however, as a project will not do well in judging if it is misplaced. Again, consult the student. Discuss the options available and the pros and cons of each. Then, insofar as possible, let the student make the final decision. A key point for consideration by both teacher and student is the question of area in which the student feels most confident for discussion of related subjects with the judges. If scoring is close, a student's knowledge of related areas may become a critical factor in final award placement.

Murphy's Law

Science fair seems to activate the well known Murphy's Law. A complete list of all of the problems and things that can go wrong at science fair would probably stretch from Canada to Mexico. Table 1 shows a few of the more common ones for which you can prepare.

* Rain on set-up day
* Hinges loosen from display boards
* Pictures and data come unglued
* A picture or chart has the wrong caption
* A misspelled word is found just before judging begins
* A critical notebook or display object was left at home or school
* The student forgot to get parent's signature on entry forms
* The parent signed the entry form in the wrong place
* Buttons fall off clothing, seams rip open, and zippers get stuck
* Batteries run down
* Critical light bulbs burn out

Table 1

Most of these and other potential disasters and problems are preventable or solvable by planning ahead. The faculty responsible at the author's school concocted what became known as the "Murphy Box." This is one of the most important items carried on the bus to science fair. Some of the items included are shown in Table 2.

* Glues: white, rubber cement, fabric cement, and epoxy
* Small traveling sewing kit with needle, thread, and an assortment of buttons
* Screw drivers: several sizes of both flat head and Phillips
* Assorted colored felt tip markers
* Portable typewriter
* Dictionary
* Major reference books related to projects which might require quick information or substantiation
* Assorted colored construction paper
* Poster boards
* Batteries
* Extension cords, preferably with fused outlets
* Utility knife
* Hammer
* Long nosed and lineman's pliers

Table 2

The reason for all the listed items except the saw should be obvious. The saw is for the project which sooner or later will show up and measure two millimeters too wide to qualify. Emergency adjustment will be necessary at set-up time to get it in the fair.

Since a corollary of Murphy's Law is that it always rains on set-up day, it is a good idea to have on hand some large plastic garbage bags to protect the displays and other materials. Students have a tendency to bring one or two individual items at a time to school during the few days preceding science fair. Require them to provide sturdy cardboard boxes for their materials.

You can avoid paper work problems by requiring that all permission forms, registration forms, protocols, etc. be in your hands at least two days before science fair day. That will give you time to review them for proper completion and signatures.

If space permits, set a deadline for completion of projects and bringing them to school a week ahead of the science fair date. During the month before this, repeat many times your usual speech about planning ahead. This will get about half of the projects there on the date specified. On that date, repeat your speech, emphasizing it with desk thumping. That will get most of the remaining projects there by three days before science fair. Then, be prepared to take whatever action your think appropriate to deal with the one or two projects which cannot possibly be completed ahead of time because of a last minute major discovery or disaster.

Some Final Thoughts

A compelling reason for requiring early completion of projects is to make certain that related disciplines have received proper attention. It may be difficult to convince some students that all the scientific knowledge in the world is valueless unless it can be communicated. Insist that reports and displays be grammatically correct. Double check for spelling errors. Make certain that the conventions for italics and capitalization in scientific names of organisms have been properly followed.

Check any mathematical applications. Be sure that statistical analyses have been appropriately selected for the type of data utilized. As preparation progresses, you may find it worthwhile to enlist the cooperation of English, math, and art teachers.

During the year, as projects begin to develop, insist that proper records be kept, preferably in a bound notebook with each page properly dated and signed. Some students will want to recopy their laboratory notebook to make it neater. Don't. This is the original record of work that was done and should be presented as such. Certainly, the notes should be legible and accurate, but judges will be suspicious (and properly so) of what they suspect might be deliberate preparations that do not really have the appearance of having been done while the work was conducted.

Table 3 on the next page is an example of a checklist which makes it easy to keep up with everything.

"KNOW THE LITERATURE!" Repeat this at least ten thousand times during the year. Insist that proper notecards be kept for bibliographic references and that these be included in the student's report in a form acceptable to scientific journals in the student's field. Counsel with the students on the importance of knowing the literature and being able to quote specific applications to the judges. This will score many points.

Finally, the big day arrives. Go in peace and with optimism and confidence. Thorough preparation and all the hard work of you, your students and their parents will pay off, if not with a trophy, certainly with the satisfaction of a job done to the best of one's ability.

Table 3

Checklist for Science Fair Project Compliance

The following must be in order before projects will be eligible for most science fairs. Be sure they are completed by the specified date.

Name_____ Project_____

Form	Date Completed	Date Returned to Teachers
Research Plan		
Parent Permission		
Safety Review		
Animal Experiment/ Chemicals		
Qualified Scientist Agreement		
Teacher/ Supervision		
IRB (Institutional Review Board) for Human experiments		
Scientific Review Committee Qualification		

Tips For Parents

Science fair can be an exciting and gratifying experience for parents. It can also be challenging and sometimes frustrating, especially for those involved the first time.

One of the most important things for your student is to plan and start well ahead of time. Students who have not participated before tend to underestimate the time required to carry out thorough experimentation. They especially underestimate the amount of time required to prepare a well thought-out and appropriate display. Presentations started the day before science fair do not get the job done. Encourage your student to start thinking about presentation as soon as the experiments are under way. Prepare rough sketches for possible display ideas. Figure 1 on the next page shows two possible arrangements of the same material in tentative display plans.

Even if you are not scientifically oriented or experienced with science fair yourself, you can offer much valuable guidance. For example, the most expensive materials are not necessarily the best. Judges like to see ingenuity in design of both the experiment and the display. This will usually rate much higher than costly "store-bought" materials. If your son or daughter lacks artistic talent, the display preparation provides a good opportunity to learn something about color and coordination of different elements in a visual presentation. Most students will have access to computers with laser or ink jet printers at school, if not at home. These should be used to make titles and text neat and attractive.

Figure 1

One of the biggest problems facing parents is a decision on the question of "how far should I go in assisting my student?" This area involves touchy decisions. Obviously, anyone will require help in handling large display boards or heavy materials. Suggestions as to arrangement of the presentation are appropriate. Basically, think in terms of, "Is physical assistance actually required to perform a task?"

If, despite all you do, a last minute mad rush for completion threatens, be prepared. You can avoid an 11:00 p.m. foray through town the night before science fair if certain standard last minute needs are on hand. Here is a short check list of supplies students frequently need at the last minute.

White   glue
Rubber   cement
Poster   board
Extra   letters
Assorted   colored   construction   paper
Plain   white   typing   paper
Labels

Section II:

Creeping, Crawling, and Flying Things

Special Notes

Experiment	**Relate To**
1. Come Into My Parlor	Math, Art

    SAFETY NOTES:
- **Be sure spider is not poisonous**
- **Avoid inhaling vapor from spray ink**

2. Dietary Supplement Effects	Medicine, Sociology

    SAFETY NOTES:
- **Avoid harmful insects**

3. Hot Sex	Math

    SAFETY NOTES:
- **Avoid Anopheles and Culex species**
- **Obtain specimens from non-polluted water**

4. Marigolds And Nematodes	Art, History, Agriculture

    SAFETY NOTES:
- **Use gloves to dig in soil**
- **Avoid human/animal pathogenic nematodes**

Introduction

The projects described in this section are unique and unusual ideas concerned with worms, arachnids, and insects. They do not duplicate common experiments found in lab manuals or many other science project books. Some of these could be legitimately put in more than one science fair classification. The decision as to proper placement should be made with consideration for your own approach and emphasis as well as the nature of the local fair.

Section II: Experiment 1

Come Into My Parlor

Description and Directions

This project is ideal for the mathematically minded bug collector or an art student taking a biology course. It will score points with math teachers as well as biology instructors. It will certainly intrigue many judges. In addition to all of that, it can be made into a strikingly unique and beautiful presentation. It is a study of the geometric design of spider webs.

The overall goal here is to study the geometry of spider's webs. This project can be approached in several different ways. A short project would be simply to make measurements of the sections of the web of one species of spider. Determine how closely the webs of different individuals of this species correlate with each other. Comparisons might include some or all of the following:

> **Shape of inner and outer sections**
> **Size of inner and outer sections**
> **Number of sections in each level**
> **Points of web attachment to supporting structures**
> **Length of web attachment to supporting structures**
> **Number of concentric levels**
> **Symmetry**

If you have sufficient math background and a significant number of webs in an experiment, statistical analysis can be used to evaluate correlation of measurements of different webs within the same species.

If time permits, this project can be extended to a comparison of web structures of different species.

Questions to Expect

We will not even think about subtitling this paragraph "Don't Get Tangled In The Judge's Web." Be certain that each spider has identification. This should include genus and species names. While it's not necessary to include such information in your formal presentation, a student should be prepared to discuss the unique features that distinguish one genus or species from others.

Spider webs are grouped and classified according to shape and means of suspension. The beginning entomologist should look this up in a good reference and be prepared to discuss it.

Beware of tricky questions from engineers. A student should be prepared to discuss comparative strengths of different geometric designs from the standpoint of both compression and stretching.

At least one judge is going to ask you if there are differences in the webs constructed by similar spiders in different locations such as predominantly sunny as compared with protected; seasonal variations, etc.

Extensions

This project could be extended to include a study of the number and kinds of prey caught in the various webs studied. Note such things as location of the prey and how it is held in the web. Note also the usual location of the owner of the web when it is at home.

Webs can be preserved in several different ways. One is to bring a large piece of construction paper or light posterboard into contact with the web. If the web is of light flimsy construction, spraying it with lacquer may help preserve its shape. Also, try spraying the web with a light spray paint or spray inks of different colors to provide a striking contrast with the mounting material. This can become an attractive part of your project display. It will also provide an easy means for making your measurements.

Figure 1.1 illustrates one way of preserving a web for study. The web was sprayed with black ink and carefully captured on a piece of white poster board. It was then covered with clear plastic and scanned into a computer. Smaller webs could be prepared in the same way, but directly reproduced on an office copier. If you have to reduce the size of the web to make your illustration manageable, be sure to measure dimensions of the original. You can then give accurate size ratios of your reproduction.

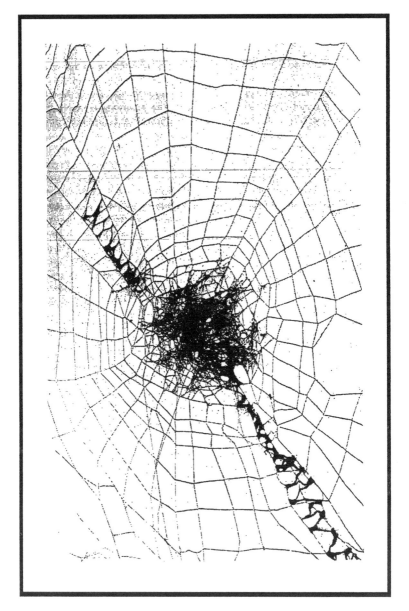

FIGURE 1.1

Section II: Experiment 2

Dietary Supplement Effects On Insect Reproduction

This is a project to determine whether or not increasing the protein content of an insect's diet will affect reproduction. Almost any insect will do, but for thoroughness, use one which undergoes complete metamorphosis and another which undergoes incomplete metamorphosis. Two insects which are common laboratory experimental organisms are *Oncopeltus fasciatus* (milkweed bug) and *Tenebrio* species (mealworm). These two insects have a rapid rate of reproduction and are easy to handle under laboratory conditions. *O. fasciatus* demonstrates incomplete metamorphosis whereas *Tenebrio* species undergoes complete metamorphosis. Food requirements for these are simple. The milkweed bug has a normal diet of shelled unsalted sunflower seed; the mealworm lives on bran meal.

Dietary protein deficiency in man and other higher animals is well known to reduce reproductive capacity. While this project will not utilize dietary deficiencies as a base, it will show whether supplemental protein contributes to reproduction.

Procedure

Adults of these insects can be found in their natural habitat or can be inexpensively obtained from school laboratory suppliers.

A small one or two gallon aquarium jar is sufficient for a large culture of either insect. Their food is simply placed in the jar, and the worms or insects placed on it. Their minimal water needs can be supplied by providing water in a small container with a cotton or cloth wick extending over the edge and down. It should not touch the food, as spoilage will result. Cover the jar with cheese cloth or screen wire. A normal range of room temperature is adequate. A good source of supplementary amino acids is casein hydrolysate. Also, consider one or more of the commercial protein supplement preparations commonly available from health food stores.

One container with the normal diet for each insect is used as a control. Prepare as many additional containers as the number of tests to be performed. Make a ten per cent (w/v) suspension of the supplement in distilled water. Saturate the sunflower seed and bran meal with the protein suspension. Allow it to remain in contact with the food for at least thirty minutes. Then dry the preparation thoroughly before placing it in the culture container. If the material is not completely dry, mold growth will occur.

Place the same number of adult insects in each container.

Leave the cultures undisturbed for two months. At the end of that time, remove all adults and immature stages and count each. Handle them gently to avoid injury. While counting, observe the immature stages for any obvious abnormalities. When you have finished counting, replace all the organisms into their original containers. Repeat the counting procedure after thirty and again after sixty days. Evaluate results compared with the controls.

Extended Project

This project could easily be extended in a number of different ways. Use more insects of different species, or different preparations of amino acid and protein diet supplements. A study of the effects of concentration levels of the supplements could be made part of any such experiment.

Another extension would be to determine the effect of the supplement on the lifespan of various insects. After the insects have reached maturity, keep the different groups separated and continue their respective diets. Look up the normal lifespan of the species you are using and compare your dietary results with the norm. This idea could be appropriate for an entirely separate project.

Section II: Experiment 3

Hot Sex

Surely you did not scan the table of contents to this book and turn to this project before looking at anything else. There is always hope, but life has its disappointments. This experiment is about mosquitoes. Sorry about that. Now, go back and read the first four chapters.

Sexual discrimination is rampant among the worms, arachnids, and insects. The female usually has the upper hand. Who ever heard of a king bee or ant? The female does not exercise her power with benevolence. She incarcerates her mate for life within her reproductive system, or she is waited upon for every need by drones. She eats the male after mating, or she sucks blood from higher animal victims in order to support her developing eggs. This project is concerned with the problem of how female numbers and female:male ratios can be influenced by such a simple thing as temperature. The absolute governing role of the chromosomes becomes variable or environmentally modified in some types of mosquitoes.

The purpose of this project is to determine the effects of temperature on the sex ratio of newly hatched mosquitoes.

Procedure

Find some still water that serves as a breeding ground for mosquitoes. Many active larvae known as wrigglers will be moving about. Use a small plankton net or other container to remove a large number from the water. Handle them gently as they are easily injured or killed at this premature stage of development. Ideally, at the same time, you should also obtain some of the water in which you found the mosquitoes. Use of it in the experiment will eliminate any questions about introducing variables by changing the water in which the mosquitoes are developing. At least five quart or larger size jars are required to conduct this experiment. All the jars should be identical. Fill each about half full of water. The specific quantity is not important, but measure to have the same amount in each jar. Prepare wire screen or cheese cloth covers which can be fastened securely over the jar tops.

Place an equal number of wrigglers in each jar. The more wrigglers available, the more valid the counts will be at the end of the experiment. It is even better if you use ten or fifteen jars where the larvae can develop at five degree differences in temperature ranging from 20 to 40°C.

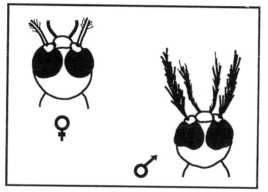

Figure 3.1

After the wrigglers have metamorphosed into their adult form, you can determine their sex. The hard way, definitely not recommended, is to place the mosquito on your skin. If it bites, it is female. An easier method is to observe placement of the eyes. Figure 3.1 shows how the male eyes are much closer together than the females. In some species, the male eyes are almost contiguous.

Figure 3.2

Two other differences are easily discerned. In most species the antennae of the male have a bushy appearance compared to the slender structure of the female (Figure 3.2). The cercus structure of the male genitalia is much more complex than that of the female (Figure 3.3).

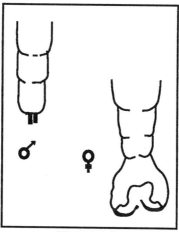

Figure 3.3

Section II: Experiment 4

Marigolds And Nematodes

For centuries many gardeners recognized that the common flowers known as marigolds (*Tagetes* species) had the power to repel nematode worms. Some planted marigolds along with their vegetable crops in areas where nematodes were a problem.

A study of the scientific literature will reveal, however, that many gardeners and scientists disagreed with the idea. Modern research has shown that the differences in results obtained in the past may have been due to species variations.

Marigolds are native to Mexico but are widely distributed in other areas. Africa is the normal habitat for *Tagetes erecta* and France is the source of *T. patula*. The characteristics of different species and cultivars of marigolds are highly variable with respect to color, height of the plant, and size of the flower.

All varieties of marigolds are easily grown. All that is required is a good garden or potting soil, warm temperature, and a bright location. Most varieties germinate readily within a week of planting the seeds. No special care is required.

This experiment is designed to establish factors which influence the anti-nematode activity of marigolds.

Procedure

The experiment will require a large planting box or it could be done in an outdoor flower bed if one is available. If you live in a nematode infested area and have outdoor planting space available, you will probably not need to inoculate the soil with additional worms. If you are conducting the experiment inside, you will need to obtain nematodes either from a soil source or from a laboratory supply dealer.

If you are using a planting container inside, it should be at least eight inches deep and a minimum of three to four feet in length and width. Fill the container with good garden or potting soil which has a rich organic content. Then, scatter your worms over the surface. They will probably quickly burrow themselves.

Obtain marigold seeds of different varieties. A thoroughly done experiment will include both African and French varieties. The original Mexican varieties, which are native to central America, do not have the form desired by growers. The popular varieties have been developed by hybridizers to obtain larger flowers.

Be sure to include in your experiment the nearest available to the original type. This is a variety which was developed specifically as a worm repellant. it is sold under the trade name **Nemagold**™. From the standpoint of worm repellant activity, it is close to and may actually surpass some of the original classic cultivars.

Plant the seeds according to the supplier's directions. After plant growth has started, periodically remove a measured quantity of soil, for example a five cm cube. Spread it out and count the nematodes present. If the marigold has repellant factors, the worms will not be found near it, but will be some distance away. Take careful measurements of the nearest distances worms are found with respect to each variety. Likewise, count the numbers per unit of soil.

With this type of experiment, it is very important to have sufficient replications and numbers for statistical validation. Obtain as much data as possible within the limits of your particular circumstances.

Your analysis of results should include many measurements of distances and numbers with correlations to marigold varieties.

This project could be extended by studying the worm repellant factors in the marigold roots. You might wish to perform an extraction of various chemical compounds and separate them by chromatography or electrophoresis procedures. Once you have separated the compounds from the plants set up a similar experiment in which you utilize the separated components without the whole plant. Again, you will need many measurements of numbers of worms and distances which they will tolerate from the plant extract.

It would be particularly important in such an extended project to have comparison of the materials extracted from different varieties of cultivars of the marigolds.

The project can be expanded as much as time and space permit.

In your project presentation, be sure to emphasize the ecological value of using natural materials in place of chemical nematocides or deterrents.

Section III:

Life Is Filled With Precious Stones

1. The Many Faces Of Uric   CHE, BIO, MED
   Acid

2. Plant Crystals   CHE, BIO, BOT

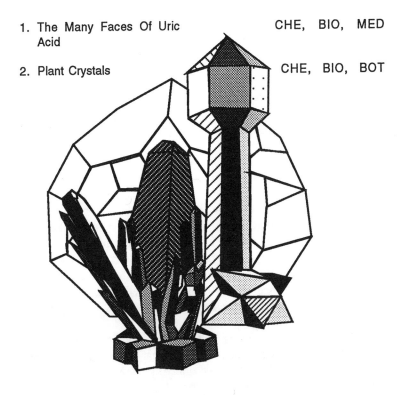

Special Notes

1. The Many Faces Of Uric Acid   EarthScience,
                                                  Physics, Art
     SAFETY NOTES:
       • **Handle all chemicals with proper precautions**

2. Plant Crystals   Earth Science,
                                                  Agriculture,
                                                  Physics, Art
     SAFETY NOTES:
       • **Handle all chemicals with proper precautions**
       • **Avoid poisonous or allergenic plants**

Introduction

Have you ever stopped to think that the world's most beautiful and precious gemstones are really nothing but chemical crystals? They are prized simply because the crystallization process inherent in them occurs rarely. In some cases, the rarest beauty occurs actually as a result of imperfections--mixed crystals. Now that we have totally removed all the romance from precious stones, let's take a look at some experiments based on crystallization.

Most people tend to associate crystals with inorganic materials, but organics also crystallize. Common table sugar is one example. There are many others in both plant and animal material. Sometimes the crystallization process *in-vivo* is abnormal and gets us into trouble. A good example of this is uric acid crystals resulting in kidney stones or gout. If the structure of a chemical compound is such that it can crystalize, there is a tendency for it to do so as the solvent in which it is dissolved is reduced. That is the basic principle behind the experiments in this section. Most of the compounds involved in these projects are in solution in their natural state.

There are a few basic principles of crystallography that you should bear in mind as you carry out these investigations. First is that the size of a crystal frequently depends upon the rate at which crystallization occurs. In general, the faster a crystal forms, the smaller it will be. Therefore, you will need to be patient and avoid too much haste in evaporating the solvents in your experiments. Another common characteristic of crystals is that the primary shape may be influenced and changed by the presence of other chemicals. The uric acid experiment will very clearly demonstrate this.

For science fair presentation of any of these experiments, you should be prepared to discuss the basic crystal systems with judges. They are sure to ask. You can find discussions of the classification of crystals in any number of text and reference books. A good starting point for a simple discussion is an encyclopedia. Then, take it from there as far as you wish to go.

Section III: Experiment 1

The Many Faces Of Uric Acid

Uric acid is formed as an intermediate product in the metabolic processes of ridding the body of waste nitrogen compounds in most mammals, including man. It is the final product for waste nitrogen excretion in birds and reptiles. Among many other unique characteristics of uric acid is its ability to form crystals of a wide variety of types and shapes. This project is concerned with investigating the conditions which are responsible for these differences.

Pure uric acid forms crystals which are almost transparent and have the characteristic shape of rhombic plates. When obtained from urine, uric acid crystals show a high degree of variation in shape and color, the result of combinations of uric acid with other components of urine. Some of the more common shapes are shown in Figure 1.

**1 . Characteristic rhombic
   plates of varying thickness**

2 . Whetstone   variations

3 . Quadrate plates

4 . Bundled forms (usually pointed)

5 . Prolonged (pointed) plates

6 . Rosette forms

7."Barrel" shapes

Experience teaches us that every good science fair project must obey
one universal unwritten rule which states that the project should
encounter one major problem. Here it is for this one: solubility of
uric acid is difficult. The compound is insoluble in alcohols, ethers,
and cold water. It is soluble only to a very limited extent (1:1800) in
boiling water. It is soluble in alkalies, concentrated sulfuric acid, some
organic bases, and alkaline carbonates. It will also dissolve in boiling
glycerol. After consideration of that, you might logically ask what
holds it in solution in urine. The answer is still theoretical, but most
biochemists think the uric acid is stabilized by the urea present.

Crystallization Procedure

This project is likely to score much higher if carried out with three different solvents for the pure uric acid. Try others to extend the project. The three most commonly used are 4 M solutions of ammonium hydroxide, potassium hydroxide, and sodium carbonate. Prepare a saturated solution of uric acid in each solvent by slowly adding the pure crystals with constant mixing until no more will dissolve.

A good initial experiment is acid precipitation of the uric acid. Place approximately 10 mL of each saturated solution in a separate test tube. To the first tube slowly add 1 M hydrochloric acid dropwise until a precipitation occurs. Repeat the process with the second tube, using 1 M nitric acid as the precipitant. Next, try precipitation with 1 M sodium dihydrogen phosphate (sodium phosphate, monobasic KH_2PO_4). Separate the precipitant from the supernatant solution and examine the crystals under a microscope. Draw or photograph the typical shapes. Use this basic procedure of precipitation for each mixture.

Procedure For Obtaining
Impure Uric Acid Crystals

Now, you are is ready study the effects of various urine components on uric acid crystal structure.

Divide the uric acid solutions into as many ten mL quantities as you wish to cope with. Determine the effects of various urine components by the addition of each separately to the uric acid solutions. An arbitrary starting quantity might be 0.01 G of the solids and 0.01 mL of the liquids.

To get a limited spectrum of classes of compounds, try the following:

Albumin
Ascorbic Acid
Citric Acid
Creatine
Cystine
Glycine

Lactic Acid
Lysine
Oxalic Acid
Purine Base
Ribose
Sodium Chloride
Sodium Sulfate
Urea
Uroporphyrins, mixed types

This is only a small portion of the chemical compounds found in urine. Oser (1965) gives a complete and detailed quantitative listing of the multitude of compounds that can be found in normal urine. To extend the experiment, use Oser's list; it provides the information needed for selection of additional materials to contaminate the uric acid solutions.

Precipitate the impure uric acid crystals with the same technique employed for the pure material. Be sure to draw or photograph each preparation. Make note of the color.

Utilize drawings or photographs to produce an attractive display for posters. It will add to the presentation if use a microscope with a micrometer scale to obtain measurements of your crystals.

At this point you have a well done and complete project. If you found it interesting and want to extend it next year, think about other components of the urine which have potential for crystallization and study the effects of various substances on those crystals. Do some research to learn more about urinary sediments and calculi. This has potential for being the start of a large and impressive extended project.

Section III: Experiment 2

Plant Crystals

Many interesting crystals can be isolated from plant materials. Except for a few microscopic cellular inclusions, the crystals to be obtained in this experiment will result from materials that were in solution in the living plant.

The experiment can be made as extensive as your time and desire permit. Many of the extensions would involve comparison of materials recovered from different plant organs such as flower, stem, and leaf. Another means of extending the experiments would be a comparison of crystals from different species of plants or from plants grown under different conditions. A very interesting project could be based on crystals isolated from healthy plants as compared with plants of the same variety suffering from nutrient deficiency, infection, or other problems.

You can expect to find both organic and inorganic materials which will crystalize from plant extracts. Consequently, the isolation methods used must cover both types. Following are some basic procedures which can be easily adapted to your specific needs.

Procedure 1
Direct isolation from
fresh plant material

Crush or grind the plant organ until at least a few drops of fluid are obtained. If you use only a small amount of plant structure such as one or two leaves or a single flower, you should expect to obtain only a few very small microscopic crystals. Larger quantities of material will yield larger and different kinds of crystals. You can obtain these larger quantities by placing the material in a kitchen food processor or blender to obtain more liquid. Filter out the solid material. Slowly, under mild conditions, evaporate the water from the liquid to obtain direct crystallization. If you start with 10 to 50 mL of liquid you will obtain both micro- and macroscopic crystals.

Procedure 2
Solvent extraction from
fresh plant material

Prepare the plant material as in the procedure above. Divide the solid debris into portions for extraction with different solvents. Most inorganic materials will be soluble from water or can be obtained by the direct method in procedure 1. Isolation of organic crystals will require different solvents for different classes of compounds. You might try a solvent of reagent alcohol initially. A more complex solvent could consist of:

reagent alcohol, 30%
propanol or isopropanol, 40%
butanol, 30%

You might also wish to try a solvent composed of higher alcohols or alkanes such as pentanol, hexanol, cyclohexanol, or heptane. **BE PARTICULAR CAREFUL IN HANDLING THESE MATERIALS AS THEY ARE HIGHLY VOLATILE AND FLAMMABLE.**

Cover the solid plant material with the solvent. Stir occasionally during the extraction period which should be a minimum of ten minutes. Remove a few drops of the liquid and place on a microscope slide. Allow it to evaporate slowly to obtain more crystals. Add more liquid to the same location and permit it to evaporate. Then examine grossly and under the microscope.

Procedure 3
For dried plant material

Crush or grind as described above. Here, of course, there will be little or no liquid directly obtained, so you will have to use extraction by solvent as described in procedure 2. Be sure in this case to use a water solvent as well as organic.

The next step for all three procedures will be to make drawings or take photographs of the crystals you obtained. Then, probably, a trip to the library will be in order to find reference materials to help identify your crystals. References on crystallography as well as some general biochemistry, inorganic chemistry, and advanced organic chemistry books will be helpful. Think about the types of compounds, both inorganic and organic, that might be present in the plant. These groups of compounds will serve as a good starting point in your comparison search.

Avoid the temptation to expect large masses of crystals from your isolations. Remember that most plants, like most animals are largely composed of water. Another large portion of the plant mass is cellulose. The true functional materials actually constitute a very small percentage of any living organism.

Your display of this project could be made most attractive with photographs or drawings of the crystals. Be prepared to answer judge's questions related to how much of the plant component exists in a crystalline as opposed to solution form *in-vivo*.

Section IV:

Plant Products

1. "Wait, Take Two Aspirin And Call Me In The Morning."

   BIO, CHE, MED, BOT

2. Color Me Like A Rainbow

   CHE,   BOT

3. Fragrances: Good, Better, Best

   CHE,   BOT

Special Notes

Experiment	Relate To
1. "Wait, Take Two Aspirin And Call Me In The Morning."	History, Sociology

SAFETY NOTES:
- **Handle all chemicals with proper precautions**
- **Avoid inhaling dust when grinding stems**

2. Color Me Like A Rainbow	Art, History

SAFETY NOTES:
- **Handle all chemicals with proper precautions**
- **Avoid poisonous or allergenic plants**
- **Avoid contact of purified dye with skin**

3. Fragrances: Good, Better, Best	Art, History, Sociology

SAFETY NOTES:
- **Handle all chemicals with proper precautions**
- **Some fragrances are irritating in concentrated form--use caution in smelling or applying to skin.**

Introduction

Since earliest recorded history, the human use of plants has gone far beyond consuming them as a primary food source. Most importantly, of course, plants provide much of the earth's oxygen. We easily recognize their uses in providing shelter and clothing. Unfortunately, they have also, all too often, been the source of violent and deadly poisons for warfare and murder. Offsetting this evil has been their use as a source of drugs for treating many ills.

One of the earliest medical records, the prescriptions of an ancient Sumerian physician recorded on a clay tablet over four thousand years ago, lists plants used to treat various diseases. It is easy to believe that ancient Egyptians, Greeks, Romans, Chinese, and Indians used plants as their drug source. Our modern minds, however, rebel somewhat at the idea of today's continued use of plant medicines. Who would believe in this world of modern chemistry and computers that about half of a doctor's prescriptions are based on drugs still obtained from plants?

Many other different types of compounds with a wide variety of uses in different industries can be extracted from plants. Recent advances in analytical chemistry have provided the means of detecting ultra-small quantities of compounds previously unknown. The experiments in this section will demonstrate extraction processes and some of the different types of products which can be obtained from plants.

You might get additional ideas as well as background information in another book by this author (Tant, 1994). **Awesome Green - The Explosive New Plant Science** is a review of new biotechnology applications

Section IV: Experiment 1

"Wait, Take Two Aspirins And Call Me In The Morning."

This project is designed to give first-hand experience with modern processes of drug extraction and purification from plant sources. It is based on an ancient--and modern--remedy for relief of fever and pain, ASPIRIN! We buy it as a simple tablet neatly packaged in a bottle. When early American Indians or ancient Greeks of 2,400 years ago needed the same kind of relief, they found a willow tree and chewed a twig or brewed a "tea" from the bark. You can isolate the active ingredient as a means of learning how today's biotechnology solves the problems of obtaining and identifying useful substances from plant tissues.

The scientific name for the active ingredient of aspirin is salicylic acid. It or its chemical precursor, salicin, is found in many plant species. Its function in the plant is not yet known. A recent study by Raskin, *et al* (1) revealed that in some lilies like the "Voodoo" high concentrations of salicylic acid caused an increase in the plant's temperature which started the production of flowers. The exact function of salicylic acid in trees is not known, but its concentration varies with the season and stage of growth.

Salicylic acid or related compounds such as salicin and salicylamide can be found in willow trees and a number of other plants. References to these are in Raskin, et al (1987) and Krieg, (1968). Aspen trees are another good source of salicylic compounds.

A chemical spot test is a means of identifying very small amounts of a substance separated from a mixture. To perform spot test, use ferric chloride to produce a visible purplish-red color to see when it reacts with salicin or salicylic acid. It will first be necessary to extract and concentrate the "aspirin" from the willow stem.

Some substances in wood will dissolve more easily in one solvent than in others. Such is the case with the compound you are seeking. Soaking the wood in alcohol for a few minutes will dissolve the salicylic acid faster than many other substances, thereby separating it

from them. Water added next will dissolve the salicin more easily.
Mixing these extracts and absorbing them into spot test paper will give
a concentrate on which to perform the identifying test.

Many organic compounds form characteristic crystals. Such is the case
with salicin. The results of the spot test can be confirmed by
comparing crystals from the extract with those from a known sample.
Figure 1.1 shows three common types of salicylic crystals.

Figure 1.1

Experimental Procedure

Chop or break up the plant material into very small pieces. The extraction will work best with the greatest possible surface area exposed. Try to obtain pieces of plant material no more than three to five millimeters in their largest dimension.

In case you want to make this procedure quantitative later, use a measured quantity of plant material at the beginning. The plant material most likely to yield a product is the stem which should be thoroughly dry before beginning. Weigh out 1.0 G and place it in a test tube.

For the first extraction, use a 50/50 mixture of methanol and reagent alcohol (ethanol). Measure the amount required to just cover the plant material in the tube. Allow this to remain for at least ten minutes. Pour off any quantity which has not been absorbed by the wood. Save this in a clean test tube.

Next, add water to just cover the wood in the original tube. Mix the water around with the wood and allow it to remain for ten minutes. Pour the remaining water into the tube with the alcohols.

While waiting for these extractions, prepare a piece of filter paper by saturating it with a 0.5% solution of ferric chloride in water. Thoroughly mix the water and alcohol extractants. Place 0.1 mL on a dry filter paper. Allow it to dry. Place another 0.1 mL in the same spot. When it has dried, place the filter paper in contact with the ferric chloride saturated paper. Observe for a few seconds. If salicylic compounds are present in the extractant, they will react with the ferric chloride to produce a purple spot where they were applied. This confirms the presence of salicylic acid. Other salicyl compounds such as salicin or salicylamide will give varying shades of reddish to brownish purple.

Every research project should have a control with known specimens. For this project, prepare a solution containing approximately 0.01% salicylic acid and a similar quantity of salicin. Test the control solutions in the same way as the plant extractant. If the results are similar, the presence of the salicylic compounds in your plant materials will be confirmed.

This is a good "quicky" project. Create an attractive presentation. Knowledge of background material and the chemistry of the procedure along with an eye-catching presentation will determine whether you win a prize or not. Another factor entering into that would be the quality of the competition. If the goal is more than this minimum, here are some suggestions for extensions:

Extended Work

Do a survey to determine the extent of the presence of salicylic compounds in different plants. This could be confined to a single plant family or based on a comparison of different families.

An interesting project can be built around a comparison of different organs and tissues of the same plant to determine the location of the salicylic compounds. For this, you would need both mature and immature tissues. Such a study could be further extended by conducting it at different times of the year. Take into account seasonal variations, changes in weather conditions, and different growth stages of the plant. While this project will stand alone with honor, it could be made into an extended project of major scope by combining one or more of the above ideas with quantitative measurement of the salicyl compounds. To do this requires a colorimeter, spectrophotometer, or other means of comparing color intensities. Colorimetric procedures would be based on the relationship between color intensity produced by the reaction product of a ferric chloride solution with different concentrations of the salicyl compounds. Procedures for establishing a reference series will vary greatly with the type of instruments available. For that reason, specific procedures are impractical here. Consult the user's guide for a particular instrument to determine the type of analysis required. If using this approach, think in terms of the project becoming one which will extend over at least a two year period.

Section IV: Experiment 2

Color Me Like A Rainbow

One of the earliest "sciences" developed by man was that of extracting colorful chemicals from plants to use as dyes for various objects. Prior to the development of the synthetic dye technology earlier in this century, plants served as the major source of coloring materials.

Be prepared for many surprises in this project. Most of them will be pleasant. A few may be ugly. The reason is that the colors extracted from leaves, flowers, or stems of plants may not be the ones which are apparent in the living material. Quite often the color seen in a plant is the result of mixtures of different colors. Table 2-1 at the end of this experiment is a list of some common plants which have served as a source of dye chemicals. Do not consider it complete; there are many many others. An interesting variation on this project would be to deliberately confine your research to plants found only in the immediate local area. The plants on this list are only a few of many referenced by Stern (1985), Adrosko (1968), and Shetky (1964).

Since the pigments are obvious, visible separation of them from the plant material is easily accomplished by paper, gel, or membrane chromatography. Here is a procedure which will serve as a good starting point.

Dye Extraction Procedure

The plant material used in any procedure should be fresh. Some pigments decompose rapidly in older tissues, particularly after they have been removed from the parent plant. Hard bark or stem material may have to be cut or chopped into small pieces. After that, it, as well as soft tissues, can be ground with a mortar and pestle or ground on a stiff wire screen.

Some dyes are water soluble. Others are insoluble in water, but will be easily extractable with alcohol or other organic solvents. To perform the extractions, place a small quantity of ground plant material in a test tube and cover it with one of the extractant solvents. To cover all possibilities use at least three different preparations. Try water, reagent alcohol, and acetone for a starter. If none of these work, the next choice would be petroleum ether or toluene. Leave the plant material in the solvent for at least one-half hour. You can judge when extraction is complete by the fact that the color ceases to become more intense with time. **Be sure to observe all appropriate safety precautions in handling these solvents.**

Almost any method of chromatographic separation will be suitable. A wide variety of methods are described in biology and chemistry laboratory manuals. Select one which can be used with available equipment. Follow the directions closely.

After achieving separation of the dye from other materials, purify it by simply cutting out the section of the chromatogram which contains it. Place this in the solvent originally used to extract the pigment. Evaporate the excess solvent under a hood to concentrate the dye. This step will conclude the experiment if the purpose was simply to obtain one or more plant-produced dyes. There are many ways to extend the experiment from this point.

Think about testing the dyes on different materials. Both papers and fabrics could be used. Another good extension would be evaluation of the effects of mordants on the fastness of the dye. You can find discussions of mordant techniques in many introductory chemistry books.

Finally, don't overlook a good application--use your own dyes to make a colorful display of your project.

Table 2.1

Common Name	Scientific Name	Dye and Source
Bamboo	*Bambusa spp.*	green dye from leaves
Black Cherry	*Prunus serotina*	red dye from bark; gray to green dyes from leaves
Black Walnut	*Juglans nigra*	brown dye from bark; brown dye from walnut hulls; red dye from rhizomes
Cocklebur	*Xanthium strumarium*	green dye from stems and leaves
Coffee	*Coffea arabica*	brown dye from roasted seeds; brown and green dyes from stems and leaves
Dogwood	*Cornus florida*	red dye from bark; purple dye from root
Fig	*Ficus carica*	green dyes from leaves and fruits
Grape	*Vitis spp.*	yellow to green dyes from leaves
Henna	*Lawsonia*	orange dye from shoots and leaves
Hickory	*Carya tomentosa*	yellow dye from bark
Indigo	*Indigofera tinctoria*	blue dyes from leaves
Lupine	*Lupinus spp.*	green dyes from flowers
Maple	*Acer spp.*	pink dye from bark
Marsh Marigold	*Caltha palustris*	yellow dye from petals
Morning Glory	*Ipomoea violacea*	gray green dye from blue flowers
Oak	*Quercus spp.*	yellow dye from bark
Onion	*Allium cepa*	reddish brown dyes from bulb scales of red onions; yellow dyes from yellow onions
Peach	*Prunis persica*	green dyes from leaves
Poke	*Phytolacca americana*	red dyes from mature fruits
Pomegranate	*Punica granatumdark*	gold dye from fruit rinds
Privet	*Ligustrum vulgare*	yellow green dye from leaves; gray dye from berries
Rhododendron	*Rhododendron spp.*	tan dyes from leaves
Saffron	*Crocus sativus*	yellow dye from stigmas
Sage	*Salvia officinalis*	yellow dye from shoots
St. John's Wort	*Hypericum spp.*	brown dyes from leaves

Section IV: Experiment 3

Fragrances: Good, Better, Best

This project will be the best smelling one at any science fair. It focuses on processes of extracting perfumes from flowers. Prepare for it by finding a large number of the flowers which you desire to use. The amount of perfume in a single flower is infinitesimally small. For example, approximately 1,000 pounds of orange blossoms are necessary for the production of one pint of orange oil.

Perfume production is one of the oldest of the chemical and botanical sciences. Perfumes are usually considered to be part of the "essential oils" of the plant. By definition, these are all compounds which volatilize with steam distillation and are extractable by means of organic solvents. Technically, some of them are not lipids. Also, be aware that many perfumes are actually mixtures of different fragrances. Some of these occur naturally; others are man-created.

Perfume production involves at least as much art as science. Almost every woman and many men are well aware that a perfume which smells pleasant to one person does not to another. Additionally, there is the factor of the final odor produced when the perfume reacts with the oils and other substances of an individual's skin. This reaction will contribute to the different effect in different persons.

Exercise caution in selecting references to cite. Be sure to obtain them from peer-reviewed scientific literature. Much material in popular writing is of questoinable scientific validity.

This project is very flexible; it could have been placed in the section on quickie projects or in the section on extended long term projects. One can make it short or involved by using different flowers, different mixtures, and different methods of extraction. Following are the basic procedures:

Procedure 1

This is the fastest and easiest method of fragrance extraction. As might be expected, it will yield a lower grade of perfume than the others.

Prepare flowers by grinding them thoroughly with a mortar and pestle. Place the ground flowers in a flask. Add a suitable organic solvent such as petroleum ether, hexane, or heptane. Close the flask and allow extraction to take place at room temperature for 24 hours.

When the extraction is complete, pour the mixture into a container with a large surface area. Set it in a well ventilated location with no open flames. Allow the solvent to evaporate. Evaporation will occur rapidly with the substances suggested above. Following evaporation, there will be an oily residue which is the finished product.

Procedure 2

This procedure is a lipid extraction. Grind the flower material with a small quantity of almost any kind of cooking fat. Old-fashioned lard is preferable because it is insoluble in alcohol. Leave the ground mixture to extract for a period of 24 to 48 hours. During this time, the fragrant portions will dissolve in the fat.

The next step is to separate the perfume from the fat. This is best accomplished by pressing the mixture through a small-pore sieve to remove flower debris. Mix the fat with an equal volume of reagent alcohol. The odoriferous material will dissolve in the alcohol. The fat is insoluble and will crystallize out on cooling. Accomplish this by placing the mixture in the refrigerator. Use a tightly closed container to prevent evaporation. The alcohol solution will now contain most of the fragrance. Terminate the experiment at this point, or further concentrate the perfume by allowing evaporation of the alcohol at room temperature. CAUTION! You could lose the perfume if it is as volatile as the alcohol.

The preparation obtained by this method should be of higher purity than that by the first procedure.

Procedure 3

This method is sometimes referred to as "cold absorption." To carry it out, you will need some glass plates and a container which can be tightly sealed.

Prepare as many glass plates as desired by coating them with lard about 3 mm thick on both sides. Layer flower petals on the surface of one plate to a thickness of about 6 mm. Place the bottom of the next plate in contact with the petals. Continue this layering and stacking procedure to obtain as much material as will be wanted.

Figure 3.1 shows a typical experimental set-up.

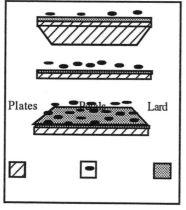

Figure 3.1

Allow the extraction to proceed at room temperature for a period of three to four days. At the end of that time, separate the plates and carefully remove the petals, disturbing the fat as little as possible. The next step is to add more petals to the same fat following the original procedure. Again, allow three to four days for extraction. Remove the petals from the fat, scrape it off into a suitable container, and add a volume of cold alcohol approximately equal to one half the volume of the fat. Mix the fat and alcohol thoroughly. Pour off the alcohol, which should now contain the perfume. If a small quantity of fat remains dissolved in the alcohol, place the mixture in a freezer for an hour or two. The fat will quickly precipitate out and can be easily removed. The alcoholic solution can now be concentrated by evaporation or used as constituted.

This procedure should yield the highest quality perfume of all methods because it will not pick up as many non-perfume substances as the other two.

Section V:

Projects For Those Who Can Take The Heat

1. Testing Household Papers   CHE, PHY

2. Change Baby's Diaper   CHE,   PHY

3. Coprolite Analysis   ESS, BIO, CHE

4. Osmosis In Biotechnology   PHY, CHE, BIO, BOT

5. The Heat Of   Motion   BIO, ZOO, MED

6. Be Fat - Be Cool?   BIO, ZOO, PHY

Special Notes

Experiment	Relate To
1. Testing Household Papers	Economics, Art

2. Change Baby's Diaper   Economics

    SAFETY NOTES:
      • **Handle chemicals with proper precautions**

3. Coprolite Analysis   Art, History

    SAFETY NOTES:
      • **Handle chemicals with proper precautions**
      • **Wear eye protection when breaking coprolites**

4. Osmosis In Biotechnology   Agriculture

    SAFETY NOTES:
      • **Handle chemicals with proper precautions**

5. The Heat Of Motion   Athletics, Physics

    SAFETY NOTES:
      • **Avoid overexertion**

6. Be Fat - Be Cool   Physics
    Safety Notes:
      • **Exercise caution with heat lamps and insulating materials**

Introduction

The name of this section was selected because the nature of the projects provides an invitation for all kinds of comments (non-scientific) and questions (scientific). The latter can become strongly oriented to physics or engineering. Learn everything possible about water pressure, diffusion, and hydraulics for Experiment 4. It will open many cans of worms for judges to play with.

Likewise, be prepared for physics questions about experiments 5 and 6. Judges are likely to ask questions about heat transfer and other aspects of heat theory.

Section V: Experiment 1

Testing Household Papers

Yes, the part of this experiment testing toilet paper will bring forth all kinds of comments. If you have a limited scientific sense of humor, conduct the experiment with paper towels and facial tissues instead. The general principles and procedures are the same.

This project can be made as long and extensive as time permits. Provided each is tested thoroughly, the more different brands used, the higher the experimenter will rate on the judge's score card. If the paper being tested is a double layered type, always conduct the test with both layers together. Following are some ideas for different tests.

Microscopic composition: Observe a piece of the paper under low magnification with a microscope. Note and photograph or sketch the cellulose fibers which are its primary composition. Observe the directions they run, their number per unit of area, and their relations to each other in placement. Note also their length and width as well as the amount of variation in size. Are they of relatively the same size, or do they differ greatly from each other?

Dry tear test: Fasten one edge of the paper securely and evenly to a support. Apply weight to the opposite side until the tissue begins to tear. Repeat this test along the opposite axis and keep careful records with respect to any differences observed vertically as compared with horizontally. Another method of conducting such a test would be to fasten the paper across the top of a can or other container as shown in Figure 1-1. Apply the weight at the center until tearing occurs.

Figure 1.1

It is easy to design tests similar to the above to determine stretching ability of the paper and elasticity. For these, stop just before tearing begins, measure changes in dimension, wait a few minutes, and remeasure to see if the paper has returned from its stretch.

Moist and wet strength tests: For these it is absolutely essential to use measured quantities of water. Repeat the tests described above, but this time carry them out on samples which have been evenly dampened for the moist strength test. Do a similar test based on wet strength. Add a measured quantity of water until the paper is thoroughly saturated. Determine the maximum amount of water to use by holding the square in a vertical position and stopping the water addition at the very moment it starts to run off the paper. Greater saturation will be achieved with very slow, even addition of the water.

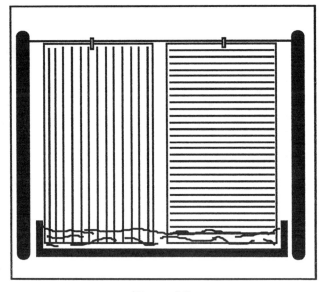

Figure 1.2

Absorbency: Determine how much water the paper will absorb by putting one edge of it into a measured quantity of water. Allow the water to saturate the square by capillary action. Take another sample of the same paper and repeat the test at a 90 degree angle from the first. See if there is a difference. This procedure can be additionally refined by determining the time for saturation to occur.

Figure 1-2 illustrates how this might be set up.

Cost effectiveness: After determining which papers perform best on the test described, calculate the cost per unit of each. Analyze this with reference to effectiveness of absorption, tear strength, etc. Determine if the least expensive paper is really the best bargain.

Section V: Experiment 2

"The Baby's Wet Again. Change It's Diaper."

The author will not touch ideas for the display of this project. This project should determine the composition and absorptive powers of different brands of disposable diapers. Start by measuring the thickness of the diaper. Then compress it and measure the compressed thickness for comparison. Make similar measurements after applying moderate stretching to the diaper.

Sacrifice a diaper to make microscopic studies of each layer. This can be done in a manner similar to that described in Experiment 1 for household papers.

Determine an average quantity of urine excreted by a young baby at one time. You can probably find this information in a reference in a library, or if not, it can be approximated by taking a freshly used diaper and squeezing the urine out of it. Use this quantity for conducting the tests where a measured amount is applicable.

Could there be a difference between absorption of urine and absorption of water? Below is a formula for making a simulated urine which is used for many testing purposes.

Whatever you do will be unique. Be sure it is also scientific. Be prepared for all kinds of comments from would-be comedians. They are guaranteed to come in all varieties--judges, teachers, parents, and your fellow students.

Simulated Urine

Component	Grams
Calcium chloride	1
Casein hydrolysate	5
Potassium hydrogen phosphate, dibasic	1.8
Purine base	0.1
Sodium chloride	12
Magnesium sulfate	0.1
Calcium sulfate	0.4
Urea	30
Uric acid	0.7
Distilled water	1200

Section V: Experiment 3

Coprolite Analysis

Coprolites are fossilized animal droppings.

If you do not live in an area where coprolites are found in abundance with other rocks, you can purchase them at reasonable prices from laboratory suppliers or rock shops. A school geology lab probably has several in its rock collections. Another source might be the geology department of a nearby college. Many "rock hound" hobbyists may have duplicates in their collections. Once a good source has been found, try to avoid overindulgence. There is an understandable temptation to attempt analyses of many different kinds of coprolites. A much better project will result if you confine your research to a detailed and thorough analysis of just one or two types.

Comparison of the physical characteristics of the coprolite with droppings of the closest related modern animal will get this project off to a good start. Get your coprolite oriented as the first step. Normally, the end which was highest in the rectum at the time of excretion will be more narrow or pointed than the other as the result of pressure from the anal sphincter muscle during the process of excretion. Next, measure the dimensions of the coprolite to compare with averages of representative modern droppings of the closest related species. This will be easiest using a mammalian specimen, particularly one which has similar modern relatives like some members of the cat group. It will be very difficult using a dinosaur coprolite.

Figure 3-1 shows a typical small mammalian coprolite from the Oligocene period. Note the crystals. The same coprolite is broken in half with a chip removed from the surface.

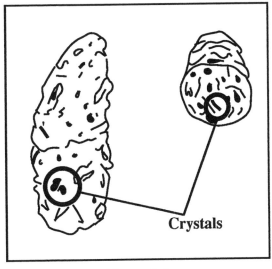

Figure 3.1

Cross examination of the specimen should include observations of color with regard to its uniformity throughout the specimen, apparent differences in mineral deposits, and differences in texture.

Next, it is time to investigate the inside of the specimen. Use a hammer or whatever means at your disposal to break the specimen open. It is best not to use another rock, to avoid contamination and mixing of its contents with your test specimen. Be sure to use safety glasses or a face shield while performing the breaking operation. Do not attempt to cut the specimen with a rock saw. The heat generated, along with the cutting fluid, could change the characteristics of your specimen.

CAUTION! Do not touch the freshly broken internal surface with your hands. Place the specimen in a plastic bag, Petri dish, or other covered plain container. Wear rubber gloves to handle it - some of the tests are sensitive enough to detect skin amino acids.

The next major problem will be to chip or grind up fragments of material from inside the specimen. Do not allow any of the outer surface material to become part of this mixture. It will be necessary to figure out a method which will work with a specific specimen. Some coprolites will be porous, soft, or brittle enough to break up easily with no more than a dissecting probe or clean screwdriver. Others may require pounding. Conduct whatever process selected on a clean, hard metal or plastic surface which has not been in contact with organic material. Be sure to wear clean rubber gloves in handling the specimen and use eye protection throughout this phase of the experiment. Divide the ground specimen into enough portions to carry out chemical tests. Some suggested methods are in the next section.

Chemical Tests

Be sure to follow all school safety regulations with respect to eye protection, aprons, gloves, etc. in carrying out these chemical tests.

Work with small quantities. For the extraction and solubilization procedures, use only enough of the solvent or extractant to cover the specimen. A glass test tube is an appropriate container. Do not stopper it tightly as gas may be generated by some procedures.

Mineral extractions: Cover the specimen with nitric acid. Allow it to remain until a large part of the rock appears to be dissolved. Several days may be required. Periodically mix the material around. When no more of the rock will dissolve, separate the solid remains from the liquid by filtration or centrifugation. If you have studied qualitative analysis in chemistry, run the liquid through the qual scheme. Another approach to identification of the minerals present would be to place a few drops of the solution on a microscope slide, allow it to evaporate, and examine the crystals which are left. These will have a shape characteristic of the chemical content. Wash the solid remains with distilled water and place several small particles on a microscope slide and examine to see if you can detect anything that looks like plant fiber, pollen grains, or other organic material. You probably will not find any in this preparation, but carry it out to be certain.

Protein And Amino Acid Analysis: Soak another portion of the specimen in distilled water for at least 48 hours. Remove the water and concentrate any dissolved matter by evaporating approximately half of the solution. This should be done at a mild temperature not over 50 degrees C. If electrophoresis equipment is available, use the concentrated specimen to test for amino acids or protein by methods recommended for the equipment you use. If such equipment is not available, do a less sensitive test by placing a drop in the center of a piece of filter paper. Allow a drop to dry and add another to the same place. Repeat this procedure several times. Then test for amino acids by spraying with ninhydrin reagent. Prepare a similar filter paper for protein analysis by the Biuret test. (Procedures for these tests are described along with others in Appendix E of this book.)

Microscopic analysis: Examine representative portions of the specimen microscopically. Look for shapes that would correlate with cellulose or pectin structures in plant material. Another item that will often show up is pollen grains. A good reference book on plant pathology or even a good college level introductory paleontology book will give pictures to use for comparison. Often it is possible to determine by comparison the family of plants which produced the pollen.

Other Stories Told

After completing a chemical and microscopic analysis, it is time to return to the original specimen for other examinations. Crack it apart again in several places and different planes if possible.

Look for different colors or other obvious characteristics which indicate the presence of different minerals. The nature of the original specimen may have caused different mineralization to occur at different times. For example, you might expect one type of rock to form in air spaces in the original excreta. Replacement and mineralization of organic matter present might require much longer, and a different type of deposit will be found in those areas. There may have been a time variation even between different organic materials. Sometimes this can be a clue to the condition of the animal's digestive system at the time the feces were produced. References in many clinical laboratory procedure manuals and veterinary manuals which will help you interpret the nature of any problems the animal might have had. They may also help identify the species by the type of feces produced.

Check with a teacher: If using present specimens to compare with the coprolites, scientific review committee approval may be required before starting your research.

The author will make no suggestions for poster presentation of this project. Questions and comments that can be anticipated from classmates and fellow contestants are also best left to the imagination.

Section V: Experiment 4

Osmosis In Biotechnology

A major problem in micropropagation of plants commercially is obtaining sterile starting material from mature plant sources. The disinfectant concentrations and times required to remove all mold spores from the plant material are very close to those which will damage the plant itself. Relatively high concentrations of disinfectants can be used on stem sections of the intact plant. When stems are cut, however, the opportunity exists for mold spores to be drawn within the xylem vessels or other tissues at the cut surface. Removal of these is difficult. One way to accomplish sterility is by drawing some of the liquids out of the plant tissues. This can be done by utilizing basic principles of osmosis and diffusion.

This experiment will demonstrate one method of removing surface contaminants from the cut ends of plant stems. It will also vividly demonstrate the effects of hypertonic solutions in removing water from the micro tubes of plant material. The osmoticums utilized are DL-mannitol and sorbitol. These are complex carbohydrate derivatives which are not normally metabolized by either plant cells or fungi. They do, however, absorb water and will demonstrate the diffusion phenomena.

The experiment can be limited to a few specimens, or it can be made extensive, comparing not only different osmoticum concentrations, but also the effects on different types of plant material such as woody stems and herbaceous stems, different sizes, and different ages of stem material. If the student is mathematically inclined, extensive calculations related to osmotic pressure functions and diffusion can be used to interpret the data. The experiment is very flexible in time requirements. It can be done on a limited basis as a "quickie" experiment, or the extensive version would provide a major project.

Procedure

For the most reliable results and significant number validity, at least
four stems of each kind should be used for each concentration of
osmoticum to be tested. The most meaningful results will include both
woody and herbaceous stems. These should be obtained immediately
before starting the experiment. Stems should be free of leaves. If
lengths of approximately two to five cm can be obtained between
leaves, results might be better. If such stem lengths are not available,
leaves should be removed immediately after the stem is cut.

It is important to avoid introducing any external artificial variables.
Try to find stems that are clean and will not have to be washed or
otherwise treated. Avoid any abrasion of the epidermal covering. When
cutting the stem to lengths, use a very sharp blade with a sawing
motion. It is most important to avoid any type of cutting which would
result in mashing and crushing the small xylem vessels. Ideal stem
length is approximately five centimeters. The experiment could be
improved and further validated by conducting it with different lengths.
Adapt the following protocol to the number and type of stems utilized
in the experiment.

1 .   Prepare solutions of DL-mannitol in distilled water
       at concentrations of 1.0, 5.0, 10.0, 15.0, and 30.0
       per cent.

2 .   Prepare similar solutions of sorbitol.

3 .   Use a control of distilled water.

4 .   Conduct the experiment with vials or small beakers
       which will support the stems in a vertical position
       without applying pressure to them.

5 .   Develop a short identification code for each stem.
       Mark the stems accordingly with a felt tip marker.
       This will not affect the results.

6 . Weigh each stem as accurately as possible immediately before starting the experiment. Record this original weight to the nearest 0.1 or 0.01 G as permitted by the balances available.

7 . Place the stems into the osmoticum with the original lower end down. It should not extend more than two to three mm above the bottom surface of the stem.

8 . After one hour has elapsed, remove the stems and place on a paper towel in a horizontal position for surface drying.

9 . As soon as they appear dry, weigh the stems accurately and record the results for comparison with the initial weights.

1 0 . If there is no weight difference, repeat the experiment extending the time of exposure in the osmoticum to periods of 2, 6, 12, and 24 hours.

1 1 . Completeness of the project can be enhanced by repeating the above procedures with the top end of the stem placed in the osmoticum.

Extended Project

The short project described above could easily be extended into a very extensive one by conducting additional experiments. Some suggested variations of the procedure might include:

1 . Compare results obtained at different temperatures.

2 . Compare results with older and younger stems of the same type.

3 . Compare results with different diameter stems of the same kind.

4 .   Compare results with different species of plants.
       This might include the use of xerophytic plants
       compared with ones having   high and constant
       water requirement.

5 .   If a microtome is available, thin sections could be
       made along the length of the stems and stained
       with any standard botanical stain to show evidence
       of tissue changes.

Section V: Experiment 5

The Heat of Motion

"Beads of sweat dotted his brow and his face was flushed with the heat of effort as he strained to move the heavy boulder."

"Her mouth was smothered by the firmness of his lips. As the warmth of their kiss became white hot, she wished it would go on forever. They parted momentarily, and then as their mouths came together again his lips blazed a trail of liquid fire across hers."

Are such statements just the literary license taken by romance novelists or is there a scientific basis to their association of motion and emotion with heat? This project will answer that question.

The experiments in this section will measure heat produced by a continuous type of muscle strain, rapid repetition of movement, and movement associated with varying degrees of emotion as engendered by kissing.

Kissing Fever

This experiment requires temperature bands or dots with a range of 30 to 40°C in one degree increments. Such temperature detectors are small flexible mylar strips containing liquid crystals of different colors corresponding to the different temperatures. They can easily be attached to and removed from the skin. Instead of, or in addition to, the temperature bands, try temperature dots made of a similar material. They are actually tiny round liquid crystal digital thermometers which can be attached to the skin in many places to map temperatures.

(Also needed is an interested and cooperative assistant.)

Procedure

Plan ahead. Prepare a list of the areas of the body topography that will be used in the evaluation of the kiss. Suggested test sites include the back of the hand, the forearm, the elbow, the neck, the cheeks, the forehead, and, of course, the lips. Kisses of different time durations should be conducted at each test site. Use a lab timer to accurately control the length of each event.

For scientific validity, as with any research, each experiment should be replicated. Allow sufficient time for the lip temperature to return to its normal ambient between tests.

You should apply the temperature detecting material to your lips and leave it in place for at least one minute before beginning the test. Have your assistant record the base temperature.

Project Presentation

Practice the presentation of this project on your friends or classmates until you can do it in a straight-faced scientific manner without giggles or guffaws. The judge will be having enough trouble.

Here are two additional hints: If any part of this project or presentation starts heating up excessively, automatic cool-down can be achieved by simply remembering that in the strictest biological sense a conventional kiss is nothing more than the prolonged anatomical juxtaposition of two *orbicularis ori*. Have sufficient background information to be able to answer judge's questions about the composition and function of liquid crystal sensing devices. Also, be prepared to discuss physiological changes associated with emotions. Many references are included in the review, Awesome Neurochemicals-The Essence of Sex (Tant, 1995).

Expanded Experiments

This experiment could be varied or extended in many different ways. Here are e few ideas to stimulate your brain into thinking about related projects.

1. Compare the surface temperature over different major muscles or muscle groups with respect to different types of exercise. For example, study the feet and legs as they respond to a static exercise such as standing on tip toes. Compare that to the results from limited movement as in weight lifting or vigorous aerobic exercise such as running or dancing.

2. Compare the reactions in different individuals. This might include a comparison of well-trained athletes with non-athletic persons.

3. Another approach might be a comparison of results in the different sexes.

Possible Trouble Spots

It is impossible to read the judge's mind, but here are a few things most will examine closely:

Environmental variables--make sure comparative tests are conducted at the same temperature and in the same air.

Time of day--avoid comparing morning tests with others done in the evening, (H'mmm, that could be another project!) Another variable here might relate to the time elapsed after consuming a meal.

Participants age--stay within a fairly narrow range.

REMEMBER--this is human experimentation. Make sure all the required approvals are obtained before you start.

Section V: Experiment 6

Be Fat--Be Cool?

It is popularly believed that many animals which live in cold environments are able to do so as a result of the insulating properties of extensive fat deposits under their skin. By modern human standards many of these species such as walruses and some whales would be considered quite obese. Likewise, many land animals such as bears may produce large fat deposits during the warm feeding months. For them, the fat may serve as an energy reserve for hibernation in addition to temperature protection.

This project is designed to show the insulating properties of fats. Since it involves the use of animal tissues, approval for vertebrate experimentation is required although the fat is not obtained from a living animal.

You should be able to obtain trimmed fat of various kinds from a butcher shop or meat department of a super market. It would be interesting to compare various sources such as beef, pork, and chicken. A variety of experimental procedures could be devised. An example is shown in the illustration below. The experimental setup shown is designed to demonstrate insulating properties with reference to heat using a source such as a heat lamp. A similar arrangement would be suitable for investigating the insulating properties against cold if ice in a plastic bag were placed directly on the upper insulation surface.

Controls for comparison might include various types of insulating materials such as fiberglass with and without backing, loose fill insulation, wood, etc. Have the thermometer in place before heating or cooling is started. Be sure to obtain an initial temperature reading and then record changes every few minutes. If you plan to use more than one thermometer be sure to calibrate them and make certain they record identical temperatures and have the same change response time.

Make you experiment more extensive and complete by comparing different thicknesses of the fats and control insulation materials. There should be an uninsulated control for each heat or cold source.

Judges are likely to ask questions about heat transfer, so be prepared by studying some good references on the subject of heating and cooling. Particularly, be prepared to discuss the "R value" which is the standard means of rating insulation effectiveness.

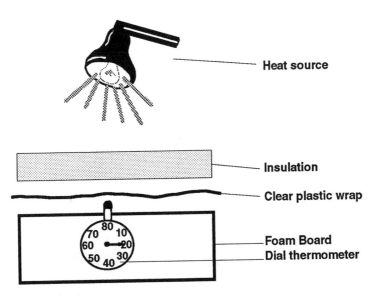

Heat source

Insulation

Clear plastic wrap

Foam Board
Dial thermometer

Section VI:

Extended, "Add To" Projects: Next Year's Big Winners

1. Androgenesis: Haploid   BOT, BIO
   Plants From Pollen

2. Bacterial Corrosion   BIO, CHE, ESS

3. Intergeneric   Protoplast   BOT, BIO
   Fusion

Special Notes

<u>Experiment</u>   Relate <u>To</u>

1. Androgenesis: Haploid Plants   Agriculture,
   From Pollen   History

   SAFETY NOTES:
   - **Handle chemicals with proper precautions**
   - **Sterilize contaminated cultures beforedisposal**
   - **Do not open cultures with mold contaminants**

2. Bacterial Corrosion   Engineering,
                                             Oceanography

   SAFETY NOTES:
   - **Handle chemicals with proper precautions**
   - **Do not use bacteria of the Genus <u>Clostridium</u>**
   - **Wear eye protection and dust mask when preparing metal specimen**
   - **Sterilize all cultures before removing metal test specimen**

3. Intergeneric Protoplast Fusion   Agriculture,
                                             Economics

   SAFETY NOTES:
   - **Handle chemicals with proper precautions**
   - **Sterilize contaminated cultures before disposal**
   - **Do not open cultures with mold contaminants**

Introduction

The name of Section VI is self-descriptive. These are projects which will extend longer than one year. They are a challenge. They are also the types of projects that are big winners in major science fairs. Be certain you are willing to pay the price in time and effort.

Section VI: Experiment 1

Androgenesis: Haploid Plants From Pollen

With normal sexual reproduction of plants, the diploid, or in some
cases, polyploid chromosome number remains constant from one
generation to the next. This occurs as a result of normal fertilization in
which one member of each pair of chromosomes comes from each
parent. As the gametes are formed, the process of meiosis results in a
reduction to one member of each pair. The fusion of the two gametes
gives opportunity for variations induced from different parents.

In many instances, it is beneficial to stabilize the genome of a plant.
This is particularly true when, for example, a single plant might show
unusually valuable characteristics as a result of mutation or some
unique combination of genes from a parent. If the required characteristic
can be obtained from one member of a pair of alleles, plants could be
reproduced vegetatively from it. All would have characteristics of the
single desirable parent. A haploid plant, of course, will be sterile. If
this is a matter of importance, the haploid characteristic can be changed
by the use of chemical treatments such as colchicine or other methods
to multiply ploidy.

An abundance of literature focuses on the subject of androgenesis.
Current research is reported frequently in such readily available journals
as *HortScience* and the *Journal of Horticultural Science*. Another
excellent but less well-known journal is *Plant Cell, Tissue, and Organ
Culture*. A brief recent summary of the subject is given by Tant and
Setzer (1989). A more detailed, but still non-technical discussion is
provided by Tant (1991). More extensive reviews along with detailed
protocols for different procedures of conducting experiments with
androgenesis are provided by Dodds and Roberts (1988), Nitsch (1981),
Pierik (1987), and Tant (1993).

The first part of this experiment is limited to obtaining cell
development from pollen. Doing so will make a good one year project
if all the variables involved are thoroughly investigated. If there are
facilities available to develop a mature haploid plant to use for various
purposes, plan on this becoming a two or three year project.
"Cookbook" projects carried out with purchased kits usually do not

score well in science fairs. However, if you are concerned with sources of some of the more unusual chemicals required, a convenient and inexpensive kit of materials for a classroom androgenesis experiment is provided by SYNTHEPHYTES™. It is available from major school laboratory suppliers.

Give serious thought to selection of the plant variety you want to use. Stay away from fancy ornamental hybrids. Many such plants are infertile. Another consideration in selection is size of the anther; large ones found on such plants as lilies are more convenient to work with. Overall, the author's laboratory has excellent results with androgenesis experiments involving common wild ("weeds") plants. These often seem hardier and may possess greater reproductive capacity than breeder-produced plants. A negative consideration for this, however, is the seasonal appearance of the flowers of most varieties. If you have a greenhouse available, or even a good sunny window, one way around that problem would be to collect seeds and plant them periodically to assure a continuing supply of plants at the flowering stage while carrying out your experiment. If you choose this approach, ascertain whether the seeds require stratification or other special treatment to achieve germination.

While walking through any field of wild flowers and observing closely, you would probably find a few individual flowers with a mutant or variant characteristic which distinguishes them from the great mass. Such plants might have valuable potential for development into a useful ornamental. The problem is to stabilize the genome and then manipulate the plant to develop its desirable traits. Androgenesis is one way to accomplish this. Development of such a variant would be a practical justification for your project in the minds of the judges.

Another consideration of great importance is obtaining the flower bud at the proper stage of development; while it is still closed, it should have no bacterial or mold contamination. This gives the assurance of starting the project without one major problem. Check the bud carefully for evidence of penetration by insects. Even the smallest hole could permit the passage of microorganisms. A bud which is too early in its development will not have pollen sufficiently mature to germinate. The following illustrations show the ideal stage of development for androgenesis.

Stages of Bud Opening

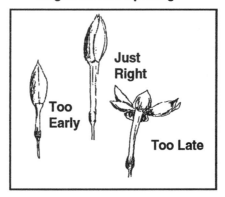

Year One: Obtaining Androgenesis

After you have selected an experimental plant, the first step will be to determine pollen viability and osmoticum requirements. A convenient procedure for doing this will be found as Experiment 4 of Section VII of this book. Next, select a medium for germination of pollen. The medium of choice for most plants is Murashige Salt Base with the addition of sucrose and mannitol concentrations which give best results in the pollen tube formation experiment. If this does not work well, try using the base medium at one half strength. Another medium which has been used by many workers is Heller Salt Base.

Most androgenesis experiments have proceeded well without any plant hormones. If any of the above does not achieve results, try them with the addition of benzyl adenine (6-benzylaminopurine) at a concentration of 1.0 mg/L.

Inoculation Procedure

Remember, the biggest problem you face is preventing mold contamination. Use good aseptic technique and work as quickly as possible. Do not leave the plant material or media uncovered any longer than is necessary. Have all your equipment and supplies in place before you start. The more movement there is in the room, the greater will be the chance of mold contamination.

CAUTION! The anther tissue itself is diploid. Be very careful to get only the pollen grains in your media. Experiment with removing the pollen from the bud before attempting to do so under aseptic conditions.

When everything is ready, follow the steps listed below.

1. Immerse the flower buds in 70% reagent alcohol for one minute.

2. Use sterile forceps to remove the buds from the alcohol and place directly into a 0.01% **Zephiran**™ solution.

3. Allow the buds to stay in the **Zephiran**™ for three to five minutes.

4. Again, using sterile forceps, remove the buds from the **Zephiran**™ and place into a bottle of sterile distilled rinse water. Gently swirl them around in it for about one minute.

5. Place the flower bud on a sterile paper towel and proceed to open it and obtain the anthers by whatever means seems to be the best.

6. Place the anthers in a sterile Petri dish or tube.

7. Scrape or shake the pollen from the anthers and place it on the sterile media in the culture tube.

8. Close the tube as quickly as possible and place it in low-level light for incubation.

9. Check the culture daily for the presence of mold contaminants. If these appear, sterilize and dispose of the tube.

10. Observe the tubes daily for the appearance of cell growth around pollen. This growth will usually occur within three to seven days. It might appear green, but more likely will be a callus-type tissue which will be yellow or brown.

As soon as the new cells have appeared, consider the experiment a total success. Different nutrients and hormones would be necessary to bring a completion of development into a new plant.

Do not expect every pollen grain to show growth. There are many factors, some inherent in the plant itself, which might prevent development. The odds of success will be greatly improved if the media is inoculated with a large amount of pollen.

Like many other aspects of the new biotechnologies, achieving successful androgenesis is something of a numbers game involving numerous replications of experimental work.

Cell development starting is a successful conclusion for this phase. It will stand as a complete project. If this has been an enjoyable experience, the next step will involve the challenge of developing a plantlet from the cell culture.

Year Two: Developing A Haploid Plant.

Different species of plants show variation in the way they develop from pollen derived cells. The most common means is for the cells to become embryogenic. If this happens, a plant can be easily induced to develop in much the same way that it would from the embryo of a seed. Since this is the most likely way to achieve success, it should be the first thing you try here.

If you are very, very lucky, the cells growing from the pollen culture will spontaneously begin to develop as embryos. In this case you need only transfer them to a medium made with the salt base previously used. Reduce the sucrose content to 3.0%. In some species, differentiation of shoots and roots appears to be hormone dependent. If you do not get results without the hormones try the following combinations:

Indoleacetic acid, 0.4 mg/L plus benzyladenine, 0.4 mg/L.
Napthaleneacetic acid, 0.02 mg/L plus benzyladenine, 1.0 mg/L.

Napthaleneacetic acid, 1.0 mg/L plus zeatin, 0.05 mg/L.

If these hormone concentrations do not achieve results, try variations including different ratios and mixtures.

It will be necessary to induce embryogenesis if your initial growth is a callus-type cell rather than embryonic. There are about as many methods of accomplishing this as there are callus cells! A method that works with one plant may not work with any others.

Sometimes, simply transferring the callus to fresh media two or three times will bring about spontaneous embryogenesis. Allow the callus to grow for tow to three weeks before attempting a transfer. If this method fails after three transfers, there is probably little hope of it succeeding.

Another simple procedure that is sometimes successful is simply to change the concentration of the media salt base. Dropping to one-half or even one-fourth the original concentration will often bring about embryogenesis.

Raising or lowering the temperature of incubation by a few degrees may also produce the desired result. Light intensity is another possible area for investigation. Some workers have reported success by using light in which the red spectrum is predominant.

Another method which works with some species is to include 2,4-D (2,4-dichlorophenoxyacetic acid) at a concentration of 0.4 to 0.5 μM in the nutrient medium. After callus growth has occurred, transfer to a similar medium without the 2,4-D will often initiate embryogenesis. Other workers have achieved good results by including benzyl adenine with a concentration of 0.1 to 0.5 μM along with varying concentrations of 2,4-D. The literature is almost unanimous in stating that the presence of cytokinins seems to inhibit embryogenesis.

If you find it necessary to get involved with a study of various hormone concentrations, a convenient protocol similar to the example here can be set up for comparing ratios. Each box of the square in Figure 1-2 would represent a tube or plate containing the mixture of hormones obtained by coming down and across. For example, the box marked X would be a tube containing 8.0 mg of hormone A with 4.0 mg of hormone B. Many will find it convenient to set up the tubes in a rack or to arrange Petri dishes to correspond with positions in the square.

Literature related to this subject will have many references which indicate that embryogenesis may be stimulated by including coconut milk and/or casein in the nutrient media. Casein is usually added at a concentration from 0.1 to 1.0 g/L of media. Coconut milk can be added at the rate of 20 mL/L. Obviously, the casein hydrolysate serves as a source of amino acids. The coconut milk serves as a source of no telling what. One theory, probably as good as any and better than others, is that coconut milk contains trace quantities of some hormone, perhaps as yet undiscovered, which stimulates embryonic development in some species. Early plant tissue culture workers used it in undefined media to achieve shoot multiplication. Think about the fact that, after all, a coconut is a seed and milk is a product and part of the endosperm.

Be prepared for a judge's onslaught if you are forced to use casein hydrolysate or coconut milk. They will home in on this like flies to a garbage can. Before they are through, you will begin to think they got together and held a secret strategy meeting to decide how to attack on this point. Their conference would have been like the one some students think teachers have when they get together to decide how they can all give a major test on the same day. The real explanation, of course, is that scientists are going to question things which are unexplained.

Figure 1.2
Hormone Ratios (mg/L)

	1	2	4	8	16	32	64
1							
2							
4							
8			X				
16							
32							
64							

Ploidy Determinations

Plan to do chromosome analysis with each different type of cell obtained. A procedure for making squash preparations of cells and two procedures for chromosome stains are listed in the appendix. Try both stain methods to determine which one works better for your cells. Be sure to read the cautions about handling the chemicals used.

If you do not already understand, be sure to learn the meanings of all terms commonly used to describe the chromosome number of cells. Most of these are in the glossary.

Another item to review before science fair is meiosis. Some judge is certain to ask you how the stages of meiosis and mitosis differ. Be prepared to discuss it in detail.

Earthquake, Fire, Flood, Tornado, And Other Disasters

Examining the chromosome stains may reveal that after two years of work to obtain a haploid plant, many of the cells are showing the 2n number of chromosomes. Is everything going down the drain--two years of work for nothing? Wait! It is not yet time to throw yourself off the nearest cliff. Perhaps this is what you are really after, and it is something that will save about six months work on the third year of this project. Just because the chromosome number is 2n, does not indicate that you have diploid cells. If the cells originated from a single pollen grain, they are what should properly be known as polyploid or dihaploid. This is simply seeing some of the practical reality of all that theory you learned in biology class about life forms having a kind of innate requirement to survive and propagate. This is just an example of the spontaneous doubling of chromosomes in a haploid cell that must be considered biologically normal. When the judges question this, be prepared to point out that the reason for a haploid plant is to use it as a tool to stabilize a genome. The haploid characteristic would have to be multiplied, anyway. In this case it is simply a great start toward final success next year.

Year Three: The Year
Of The Plant

The mission this year, should you decide to proceed, is to get a di- or poly- haploid plant from your haploid or polyhaploid cells. This may prove to be the easiest part of your whole project, especially if you are starting with callus or embryo cells which have already spontaneously multiplied their chromosome number.

Start thinking now about presentation. This could be your big year-- the year you take it all home from science fair. There is nothing like a plant in a pot to be examined by the judge. Prepare to have a normal "wild" plant of the chosen variety for comparison with the product of this research.

Plantlet Development

Organogenesis is the proper term to describe the initial stages of plant development from an embryo. The embryo itself is not considered to be an organ.

An early worker in this field, Torrey, (1966) was the first to suggest that organogenesis actually is a second stage in callus plant development. He observed that the callus cells underwent changes which gave them more of a meristem cell characteristic. Other researchers, including Street (1977) and Thorpe, *et al* (1982) confirmed Torrey's hypothesis. It is interesting that these workers found the meristematic region developing most actively at the interface of the cell mass and medium surface when agar was employed.

If meristem development is actually the process involved, organogenesis is more easily explained. In some cases, plantlets will begin to develop spontaneously from meristematic or embryonic cells without the addition of growth regulators. The more usual process involves the addition of plant hormones. As would be expected, a high ratio of auxin to cytokinin tends to stimulate root development; a lower ratio favors shoot development.

Procedure For Plantlet Initiation

If there are any references in the scientific literature about tissue culture production of the chosen plant, the media recommended for it would be a good starting point. If not, there are three good basic media listed in the appendix of this book. Murashige A is more suitable for herbaceous plants than the other two. Murashige B is used with many woody plants as well as herbaceous. All three of these media will require the addition of sucrose at a concentration of 2.0-3.0%. If you wish a solid medium, plant tissue culture agar at a concentration of 8.0-12.0% may be employed. Some workers have suggested that the addition of thiamine at a concentration ranging from 0.01-0.03 mg/L is beneficial.

Proper inoculum quantity for plantlet initiation is often a hotly debated topic among tissue culturists. If you are using a callus or embryogenic cell culture, your best results will probably be obtained from an inoculum at least two or three millimeters in diameter. Be sure to get some cells which were in contact with the original medium. If the source culture has large pear shaped embryoids, it might be possible to start a plantlet from these individually. This would be by far the most desirable state affairs because you would know exactly the origin of your plant.

Plan to start this project immediately after the current science fair. A lot of time may be needed. For example, three to four weeks may be required for a plantlet to start to developing from a cell culture. Do not discard the cultures before at least four weeks have elapsed. You may need to try several different media and hormone ratios to achieve the desired result.

The culture may show root, but no shoot development. If such is the case, increase the cytokinin to auxin ratio. On the other hand, if there is shoot development without roots, the addition of auxins such a IBA will bring about root formation. In either case, let the mass of tissue grow until it begins to crowd its space. This will enable you to divide it and multiply your culture as it is transferred to new media. Best results will be obtained if the divisions contain at least two or three shoots or major roots, as the case may be.

Once a plantlet has started to develop, make sure that the culture is where it can get plenty of light. This will enable the plant to start its own process of photosynthesis and create its own needs. Usually a distance of 8-12 inches from two 48-inch fluorescent lights is sufficient.

The Final Step

An *in-vitro* plant has now been produced. The next problem is to adapt it to normal growth conditions.

The plant will need about two weeks to adjust to a normal life on its own. Start the process by carefully removing the plant from its nutrient, rinsing the roots with sterile water, and planting it in small pot of sterile potting soil. A fine mix is best.

For the first few days your plant will be highly sensitive to drying, since most *in-vitro* plants do not form a cuticle on the leaf surface. Help it adapt by keeping the container covered. Do not allow the cover to touch the plant. Keep the plant out of direct sunlight. Make sure it has plenty of water. Misting it two or three times a day with sterile water may be helpful, but be sure the leaves dry off before recovering it. Gradually increase the period of time the plant is not covered. Work this out so that by the end of two weeks the plant will be growing on its own without protection. After the plant is well established, make the final chromosome analysis. If ploidy has not spontaneously multiplied, treatment with colchicine may be required. The chemical can be added to the soil or a very dilute solution sprayed on the leaves. Handle colchicine with care. High concentrations of it can be very toxic. After two or three weeks recheck for ploidy. Regardless of how this turns out, you now have a final completed project.

Take plenty of time to prepare your posters and other aspects of your presentation. This is not the time to let a first class bit of research fail to get the attention it deserves. Good luck!

Section VI: Experiment 2

Bacterial Corrosion

You will find about as many variations of a definition of corrosion as there are books on the subject. Essentially, they all boil down to about the same thing. Corrosion is the term applied to changes in metallic objects brought about by chemical or other environmental influences. The term corrosion usually implies that these changes are unfavorable. Corrosion problems affect many industries. Some of the problems occur in surprising places. It is difficult to obtain figures relative to the exact cost of corrosion to industry; however, estimates put it at billions of dollars per year in the United States alone. Such an amount of dollar loss means, of course, that there are many practical applications of the processes investigated in this experiment. Keep that in mind to emphasize to judges.

Normally, we tend to associate corrosion with oxidative processes which occur in an aerobic environment. Such is the case with common chemical corrosion. Another type of corrosion, however, is associated with bacteria. Some of these are aerobic organisms producing corrosion by action of their metabolic byproducts on metal. A surprising amount is due to the actions of anerobes living in the complete absence of oxygen.

There are indications in some of the earlier literature dating back to the late 1800's, that a few bacteriologists suspected the involvement of some aerobic organisms such as those now placed in the genera *Sphaerotilus* and *Gallionella*. They did not specifically define the processes, nor did they associate particular species with them. Another vague observation around 1900 was the implication of acid production by members of the genus *Thiobacillus*. Proof of the role of anaerobic bacteria in producing corrosion under non-oxidative conditions did not occur until 1934, when von Wolzogen Kuhr and van der Vlught made an historic discovery implicating sulfate reducing bacteria as the etiologic agent. Over the next quarter century, many workers conducted investigations which elucidated the processes involved. Some of the outstanding research of historical interest was conducted by Starkey and Wight (1945) and Linstead (1947). The renowned marine microbiologist, Claude E. ZoBell, summarized the then existing knowledge about such organisms in his monograph in 1946. During the 1950's, E.O. Bennet and his graduate students began to apply some of the basic knowledge that had accumulated by then. They were attempting to solve or prevent the problems of anerobic bacterial corrosion. If you can locate a copy of his long out of print, **Petroleum Microbiology**, you will find an excellent review of the subject by Ernest Beerstecher, Jr. in 1954.

The term "pitted corrosion" often applied to the results of bacterial processes is very descriptive of the difference between bacterial and common chemical corrosion processes. The bacteria, growing in small colonies on the metal surface, tend to produce small localized, sometimes almost microscopic pits in the metal surface in contrast to a more generalized type of attack in chemical processes. When the surface of a piece of affected metal has been cleaned, even the small pits are clearly visible under oblique light.

This project requires advance planning and starting early--very, very early. In fact, it probably should be described as a project for next year.

The simple fact which you can do nothing to change is that bacterial corrosion processes are very slow. After you have established a culture of the bacteria with metal test specimens, you will not even begin to see results for a period of at least three months. From the standpoint of a successful science fair project, the time involved is not necessarily bad. The fact that you knowingly initiated such a project and had the patience and determination to conduct it over the year or longer required, will score very favorably with most judges.

Most bacterial corrosion occurring in an anerobic aqueous environment is the result of action by members of the genus *Desulfovibrio*, which are strict chemoautotrophic anerobes. In environments containing a higher quantity of organic matter, some members of the genus *Clostridium* are capable of producing hydrogen sulfide byproducts of their metabolic processes. You would do well to avoid the latter as many members of the *Clostridium* group are pathogens, producing harmful or even fatal toxins. No pathogenicity problems have ever been ascribed to the *Desulfovibrio*. *D. desufuricans* and *D. aestuarii* are the species most commonly involved.

Procedure

All experiments should be conducted in screw cap test tubes, minimum size 16 X 100 mm, or other glass containers which can be filled to the top and tightly sealed with a plastic lined screw cap. Since you will be working with strict anerobes, all oxygen must be excluded. Among other things, this means **avoid the use of polystyrene disposable culture vessels** because polystyrene does permit slow diffusion of oxygen. Standard flint glass prescription bottles with screw caps are good containers for larger specimens than culture tubes permit. In all cases, prepare the media with the specimen in it by filling the culture vessel to the rim and placing the cap on loosely for autoclaving. The process of sterilization with steam under pressure will result in removal of all oxygen in the media. As soon as pressure comes down, the media should be immediately removed from the sterilizer and the caps screwed on tightly to prevent re-entry of oxygen as the solutions cool. During the cooling process, the tubes or bottles should be covered with a pan or other protection to prevent flying glass should implosion occur. This is not likely to happen with good quality glassware but the possibility does exist, so take precautions.

Media and Inoculum Preparation

A survey of the earlier literature will quickly reveal that each investigator concocted his own favorite nutrient medium for growing sulfate reducing bacteria. The proliferation of different media made comparison of results from different laboratories difficult, if not meaningless. In an effort to rectify this situation during the 1960's, the American Petroleum Institute (API) established a subcommittee to recommend standards for use in the petroleum industry. One outcome of this committee's work was the medium now known as API Sulfate Reducer Broth. It is easily prepared and will give consistent results with most *Desulfovibrio* species. It is even suitable for the cultivation of halophilic variants by inclusion of an appropriate sodium chloride content to correspond to their osmotic requirements. Table 1 lists the formulation for one liter of the API Broth.

Table 1
Formula - Sulfate API Broth Ingredients per liter:

Yeast Extract	1.00G
Ascorbic Acid	0.10G
Sodium Lactate	5.20G
Magnesium Sulfate	0.20G
Dipotassium Phosphate	0.01G
Ferrous Ammonium Sufate	0.10G
Sodium Chloride	10.00G
Distilled Water To Make	1.00L

Analytical grade chemicals should be used. Prepare the medium by dissolving the various components in approximately 900 mL of distilled water. Be sure to add the ingredients and dissolve each in the order listed. When all are dissolved, add distilled water to make a total volume of 1.0 liter. Following autoclaving, a slight precipitation will be noted. This is normal.

If your school does not maintain a stock culture of the *Desulfovibrio* species you wish to use, you can obtain one from many laboratory supply companies or from the American Type Culture Collection. Another good source might be a nearby university laboratory. Many private industrial laboratories maintain stock cultures, but for reasons of company policy may be prohibited from supplying a culture.

Your first step will be to establish an adequate supply of bacteria. You should prepare at least 20 tubes for growing your stock cultures from the initial culture. Assuming that it is obtained in a broth form, each tube of new media should be inoculated with approximately 0.5 mL of the broth. Be sure to mix it by inversion before removing the inoculum. *Desulfovibrio* species have a tendency to grow with their greatest concentration in the precipitate at the bottom of the tube. Now, **CAUTION**. Be very careful in all manipulations to avoid mixing air with the contents of the tubes. Oxygen is toxic to many varieties of sulfate reducers!

If you have sufficient inoculum for your subcultures, growth will become apparent within five to seven days. The appearance of the medium will make detection of the logarithmic growth phase easy. As the bacteria grow they release hydrogen sulfide, which combines with the iron in the medium to produce an insoluble black iron sulfide precipitate. The chemical reactions of the reduction of sulfates to sulfides is, of course, the basis for the bacterial corrosion of metals. As the hydrogen sulfide is produced, it will react with iron or aluminum ions to produce the metallic sulfide. Removal of the metal ions from the surface is the process of anaerobic bacterial corrosion.

Once you have obtained a sufficient supply of stock culture, you are ready to proceed with the experiment.

Obtain thin pieces of the different kinds of steel and/or aluminum you wish to test. Cut these into strips of a width which will readily go through the mouth of your culture vessel. Their length should be about half the depth of the media in the container. Try to get all specimens cut to a uniform size. Thoroughly clean each with a strong detergent solution and brush to remove oil and other particles which might be adhered. Rinse thoroughly with distilled water. Use a fine sand paper to polish the specimen to a smooth surface. Rinse in distilled water again and dry with a lint free cloth. Weigh each specimen and record its initial mass. Use an analytic balance which will weigh with accuracy to at least one mg. Have your tubes filled with media and numbered before you start weighing the metal specimens. As you complete the weighing of each, place it immediately in its tube. This will prevent variables induced by contact with moisture in the air. It also provides a convenient means of keeping up with the weight of each specimen. As soon as the specimen preparation is completed, autoclave the media with the specimens immediately. Doing so will prevent any effects occurring from dissolved oxygen.

After sterilization is complete, inoculate each tube with an identical quantity of your culture as described above. The optimum growth temperature for most *Desulfovibrio* species is in the range of 20 to 30°C. This means that they will normally do well at room temperature. Slightly faster growth will be obtained at the upper end of this range.

Allow the cultures to incubate with the metal specimens for at least three months.

At the end of the incubation period, loosen the caps slightly and sterilize the tubes by autoclaving before removing the metal specimens. After removal, wash off the specimens with a detergent solution, dry, and weigh. Be sure that all particles have been removed from the surface before attempting to weigh the specimen. If necessary, use a small soft brush to facilitate removal, but be careful not to scrape the metal itself. Be sure that all specimens are carried through the same cleaning process whether there is evidence of material adhered to them or not. Doing so will prevent one variable in your results. After weighing, calculate the percentage of weight lost and examine the specimen, using a magnifying glass or binocular microscope to find evidence of pitting.

You have now established evidence of bacterial corrosion. Certainly, you included an uninoculated control in your initial set-up! If you failed to include the control, you might as well start over again and plan to use the project for year after next instead of next year. If you are not going to be around by that time, pass on your wisdom to a younger student and find another project for next year.

"What Do I Do With The Results"

"So I demonstrated anerobic bacterial corrosion of metal. So what?" That might be the reaction of some science fair judges as well as yours. You should have given some thought to applications and comparisons before you started. If you did, the foregoing question would not be a problem.

Good comparisons could be based on different types of metal. Another excellent question might relate to whether there is a difference between cultures grown under static conditions, absolutely still during the incubation period, as compared with similar cultures shaken vigorously every day or two during incubation.

A good extension or expansion of the project would be to include a study of the effects of various corrosion inhibitors. A trip to the library or talking to corrosion engineers will reveal a large number of compounds which have potential as corrosion inhibitors. Different classes of these might be compared for effectiveness in your particular test system. You will score highly with judges if you can obtain some cost figures for the various processes and present data related to cost effectiveness of different materials.

Possible Pitfalls

Oops, sorry about that. It was not intentional. If you wonder what that statement was all about, stop and think a minute about that subheading in this project! Now let's get serious.

First of all, remember that most science fairs have strict rules requiring prior protocol approval for projects dealing with bacteria. Be sure that your paperwork is in order before you begin.

Judges are rightfully going to question the amount of inoculum in each container. It is very difficult to do anerobic plate counts of sulfate reducing bacteria with the equipment available in most school laboratories. You should make the judge aware that you realize the need for such data. One way to do this is to emphasize thorough mixing of the inoculum before use. Another test which is not really very good, but is better than nothing, might be a direct microscopic count of the bacteria in the inoculum. For this you will need a ruled slide. One of the best is the type used for counting bacteria in milk. Follow the manufacturer's directions for obtaining counts with a specific kind of slide.

Be certain that you have avoided any variables in preparing and weighing the metal test specimens. This includes cleaning procedures as well as time of exposure to air in handling. The latter would be especially important if you live in a highly humid climate.

Other questions judges are likely to ask relate to the effects of other types of bacteria in the same water. Very few natural water sources contain only *Desulfrovibrio* species. Would the corrosion effects be the same in a mixed culture? One answer to this can be found in some of the much earlier literature. As early as the 1950's, Bennet and Wheeler and Stelzner and Bennet demonstrated that anerobic bacterial corrosion can occur even in oxygenated environments. If anerobic bacteria, particularly those which produce a mucoid capsule, are growing with the sulfate reducers on the metal surface, the anerobes will utilize the oxygen, producing localized anaerobic conditions in which the sulfate reducers can thrive. You will want to point out to the judge that you are aware of this, but that it actually might constitute an entirely separate project.

Since you are working with a system in which at best there are many possibilities of uncontrolled variables, each experiment should be carried out in duplicate. You will make even more points on the judge's score card if you can repeat the experiment at a different time. It would not be necessary to wait until the first series is completed before starting another.

You can head off questions related to significant differences between trials by applying appropriate statistical methods to validate the data obtained. Be prepared to discuss whether the differences could be the result of random chance in contrast to other factors.

Judges will consider this project an open invitation to pick your brain apart to find out what it contains about oxidation - reduction reactions.

Resurrect that Chemistry I textbook you laid to rest last year!

In the unlikely event you come up with no significant differences between different metals, incubation conditions, etc., be prepared to make that a positive, not negative aspect of your report. You still have a highly successful project!

Section VI: Experiment 3

Intergeneric Protoplast Fusion

This experiment relates to one of the most advanced technologies in biology.

Protoplasts are all the contents of a cell inside its wall. They are prepared by removing the cell wall. Since the cell wall is made of materials such as cellulose and pectin, which are different from other cell materials, it is fairly easy to remove the wall. The simplest way to do this is to treat the cells with enzymes which will digest the cellulose and pectin, but have no effect on the protoplast.

Isolated protoplasts are being used for many types of plant research. One major area is in studies of movement of materials through the cell membrane.   Another is the subject of this lab--fusing together protoplasts from unrelated plants to produce a hybrid which could not otherwise be obtained. For example, such a cross has been achieved between a tomato and a potato. It is called a pomato. While it did not have the combination of traits its makers had hoped for, it represented a major advance--after all, you learn to walk before you run.

The conventional way of producing hybrids has been to take the pollen from one variety of plant and place it on the flower of another. Many of our economically valuable crop and ornamental plants have been made in this way. Two problems exist with it, however: 1. Time--often years are required to achieve a successful cross, and 2. These hybrids are limited to closely related varieties within a single species.

Asexual fusion of cells from unrelated plants holds great promise for the development of new species in a way that will overcome the older problems. The first phase of such a process is what you will be doing in this lab. It involves fusing protoplasts from two different unrelated plants. It is followed by organization of a cell wall and growth of the new hybrid cell. The second phase, which is beyond the scope of this kit, is the most difficult and time consuming.   It involves multiplication of the new cells and production of a plant from them.

After protoplasts are obtained and suspended in a liquid, a few fuse without any help. The rate of this natural fusion can be greatly increased by carrying it out in a solution of polyethyleneglycol (PEG). Since the fusion occurs at random, many protoplasts of the same kind will fuse, producing homokaryons. The desired heterokaryons from the fusion of unrelated protoplasts can be easily detected in this experiment because of the different appearance of the two kinds of plants used.

Material Required

First Day:

Disposable dropper pipets
Enzyme solution 1 (Cellulase)
Enzyme solution 2 (Pectinase)
Sterile disposable culture tube with cap
Microscope slides
Cover slips
Carrot cell culture
Tobacco cell culture

Second Day:

60-mesh sieve
Centrifuge tube
Disposable dropper pipets
Polyethyleneglycol solution (PEG)
Microscope slide
Cover slips
Wash solution
Protoplast-enzyme mixture prepared first day

Procedure

These steps are summarized in the flowchart on page 134.

First Day:

1.   Use a disposable pipet to place 2.0 mL of mixed enzyme into one tube of carrot callus.

2.   Use a new pipet to place 2.0 mL of mixed enzyme solution into one tube of tobacco callus.

3.   Cap the culture tubes and mix cell suspensions thoroughly, but gently with the enzyme solution.

4.   Place one drop of each cell suspension on a microscope slide, cover with a cover slip, and save to examine.

5.   Use low, then high power on your microscope to examine the cells on the slides you have made. Note particularly the shape of the cells and the presence of the cell walls. Sketch each on your lab report and make note of any differences which will help you distinguish the carrot from the tobacco.

6.   Allow the tubes to remain at room temperature for approximately 24 hours.

Second Day:

By this time most of the cell walls should have been digested away, leaving only the protoplasts and some debris. Compared to intact cells, protoplasts are very fragile, so all mixing, pouring, etc. for today's work should be done as gently as possible.

TIME IS CRITICAL FOR SOME OF THE STEPS THAT FOLLOW. YOUR SUCCESS RATE (they may translate into having first place or nothing for this project) MAY DEPEND UPON HAVING EVERYTHING READY AND KNOWING EXACTLY WHAT COMES NEXT <u>BEFORE</u>

YOU START. Aside from that, who wants to do all this work and then miss seeing the result as it happens? So.....read the rest of this and have everything ready before you start!

1.   Place about 3.0 mL of each protoplast enzyme solution into a syringe filter.

2.   Push the mixture through the 60-mesh sieve into a centrifuge tube. Discard the material that remains on the sieve.

3.   Place the tube in a centrifuge and spin it at low speed (100 to 400 RPM) for 30 seconds or until the protoplasts have formed a distinct pellet at the bottom.

4.   Use a pipet to carefully remove and discard the liquid supernatant solution.

5.   Add approximately 5.0 mL of wash solution to the pellet and shake the tube gently to suspend the protoplasts.

6.   Again centrifuge at low speed until a pellet forms at the bottom of the tube.

7.   Remove and discard the wash solution.

8.   Remove the protoplasts from the tube and deposit then in the center of a Petri dish.

9.   Add about 20 drops of polyethyleneglycol solution (PEG) to the protoplasts and carefully stir the mixture with the same pipet used to add the PEG.

10.   Place a drop of the mixture on a microscope slide, cover, and quickly observe under the microscope.

11.   In about 30 to 120 seconds you will observe some
       fusion occurring. Since the fusion occurs between cells at
       random, many of the fusions will be between the same kinds
       of cells. Look for fusions between the two different kinds.
       Select a few good fields and observe for ten to fifteen
       minutes. Sketch the appearance of fused cells
       (heterokaryons) at the beginning and at the end of the
       observation.

During the next 24 to 48 hours the fused protoplasts will undergo
internal reorganization, and if successful, will begin to form a cell wall
to complete the formation of a new cell. If the cell has succeeded in its
reorganization, mitosis will occur, followed by cell division. In an
intergeneric fusion which you have just accomplished, the odds are
greatly against further cell growth because of chromosome number of
differences, etc. under these conditions, hundreds of fusions may be
required to increase chances for success in obtaining further reproduction
and development.

Procedural Notes

You can easily make a convenient filter for the wire mesh step by
cutting a circle of the mesh to the exact inside diameter of a 10-20 mL
plastic syringe. Push it all the way in so that it rests on the shoulder
of the barrel.

Do not feel restricted to carrot and tobacco cells. They are used in the
directions only because they are readily available from laboratory
suppliers and because they are easy to distinguish from each other. If
you use other plants, be sure to select types you can easily tell apart.

Project Extension

If you achieve successful fusion and want to extend your project another
year, you will find directions for cell multiplication and plant initiation
in several of the books about plant biotechnology listed in the
references.

Flowchart for Protoplast Fusion

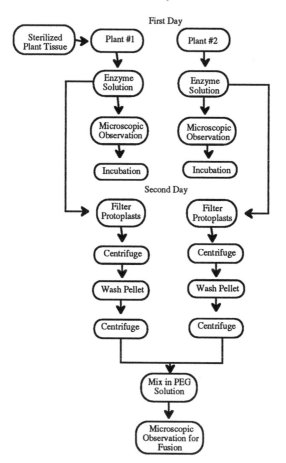

Section VII:

Quickies For Those Who Waited Too Long

1. Cobalt: Bacteria, B-12,   BIO,   ESS
   an Purple Pigment

2. Bacteria, Pot Holes And   BIO, CHE
   Busted Axles

3. Cellulase Digestion Of Plant   BOT, BIO
   Cell Walls

4. Conditions Affecting   BOT, BIO
   PollenTube Formation

5. Eat Chocolate Before You   BIO, MED
   Go To Bed?

6. Calling Dr. Ape   BIO, ZOO, MED

7. Color Your Brain   BIO, MED

Special Notes

Experiment   Relate To

1. Cobalt: Bacteria, B-12,   Art, Economics,
   And Purple Pigment   Medicine

   SAFETY NOTES:
   - **Handle chemicals with proper precautions**
   - **Use only non-pathogenic bacteria**

2. Bacteria, Pot Holes And Busted Axles   Economics,
                                              Engineering,
                                              Earth Science

   SAFETY NOTES:
   - **Good bacteriology requires handling all unidentified cultures as if they were pathogens.**
   - **Avoid inhaling asphalt and H_2S vapors**
   - **Watch traffic when obtaining road samples**

3. Cellulase Digestion Of Plant Cell Walls   Agriculture,
                                              Economics

   SAFETY NOTES:
   - **Handle all chemicals with proper precautions**
   - **Sterilize contaminated cultures before disposal**
   - **Do not open cultures with mold contaminants**

4. Conditions Affecting Pollen Tube Formation Agriculture,
                                              Economics,
                                              Entomology
   SAFETY NOTES:
   - **Handle chemicals with proper precautions**
   - **Sterilize contaminated cultures before disposal**
   - **Do not open cultures with mold contaminants**

5. Eat Chocolate Before You Go To Bed?   History,
                                                Food Science

    SAFETY NOTES:
- **Use eye and hand protection in glass bending**
- **Avoid use of potential pathogens**
- **Sterilize all cultures before disposal**

Introduction

During a misguided moment of enthusiasm last October, probably brought on by temporary brain dysfunction, you promised to do a science fair project. Suddenly, it is the middle of February, and the teacher is asking about progress with the projects because science fair is just four weeks away. Creeping forebodings of panic suddenly start racing. What to do?

The projects in this section can get you out of trouble. While they can be done quickly, they are not nominal projects that will be embarrassing unless you throw them together.

These ideas do not require much in actual experimental time. That free time should be spent in the library. Check out all the books you can carry and stay up late at night learning background and related information about their subject. That will insure a good presentation of the project. If you become quickly knowledgeable about the subjects, they have potential for winning awards.

Many of these projects are excellent for first-time science fair participants. They will give you a chance to see if you like science fair work without sacrificing too much time. After going to the first science fair, many will probably become enthusiastic about it and want to do more. The ideas in this section can easily be extended into major projects next year. If you think you might want to do this, be sure to include some of the ideas in the future work section of your display.

Section VII: Experiment 1

Cobalt: Bacteria, B$_{12}$, and Purple Pigment

The element cobalt is widely distributed in nature. Geologists estimate that it is the thirty-third most common element in the earth's crust. However, the concentrations of cobalt at single sites are relatively low. The difficulty in obtaining cobalt in large quantities accounts for its high price.

The Cobalt Cycle In Nature

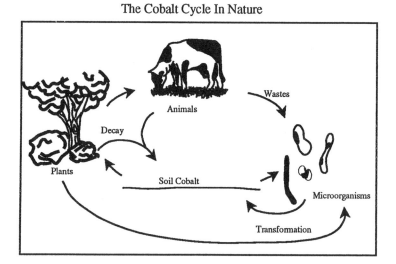

Cobalt is an essential element for most forms of life. It exists usually as part of vitamin B$_{12}$. Because of the uneven distribution of cobalt in the earth's soils, there are many areas with cobalt deficiencies which results in anemias in both man and lower animals. Figure 7.1 shows the cobalt cycle in nature.

Young, in 1956 and 1960, studied major soils of the world and found that the cobalt content of most is limited to a few parts per million of the total soil mass. Conventional processes of mineral concentration by leaching from rock do not work well with cobalt because of its interrelationships with other elements. This is particularly true with cobalt in its original forms in igneous rocks. Much of the cobalt in sedimentary material has undergone chemical reactions to exist as sulfide and other complexes which are subject to further change by bacteria, or in some cases, direct absorption by plants.

Many different cobalt-microorganism relationships can be studied. Any one of the procedures described in the following sections would make a good science project. Combining two or more of them could produce an outstanding continuing project for two to four years.

Table 1-2
Organisms For Metabolism Studies

Bacillus natto
Lactobacillus lactis
Pseudomonas calleyecolor
Streptococcus fecalis
Streptomyces griseus
Thiobacillus thiooxidans
Torula utilis

The use of microorganisms as a means of biochemical prospecting for cobalt has been very limited. This is an area in which exploration might be unique and useful. Many plants are adaptable for this purpose. One of the highest contents of cobalt reported is a concentration of 661 parts per million in the tree *Clethra barbinerias* in Japan. Yamagota and Murekia (1958) showed that the average cobalt content of the tree could be as low as 5.3 ppm, correlating closely with the soil content of cobalt. Other plants known to accumulate large quantities of cobalt include various species of birch, balsam, and white pine trees, along with blueberries.

Allow plenty of time for projects in this section. Most of them will require preliminary work to adapt the general procedures given to specific combinations of microbes, soil, etc. While adding to the difficulty, this will also add to the total on the judge's score card.

Procedures

Chemical Analysis For Cobalt

Absolutely essential for any of the project ideas described will be chemical analyses for cobalt. While a qualitative procedure could be used for simpler projects, quantitative analysis is necessary for a first class job. Many tests for cobalt are available. One good source might be some of the different procedures used in Chemistry II or AP Chemistry courses. Most instruction and experiment manuals for different types of spectrophotometers and colorimeters will have quantitative procedures for cobalt analysis. The one recommended by a particular manufacturer is usually ideal for the instrument described. If instrumentation is not available, simple-qualitative tests can be made by comparison of the unknown amount of cobalt with various known concentrations.

Two different identifying reagents are commonly used for the detection of cobalt in chromatography and electrophoresis procedures. One of these involves the use of 8-hydroxyquinoline (oxine). It may be used in one of two different ways. The first is simply to prepare a 1.0 per cent (v:v) solution of the 8-hydroxyquinoline in concentrated ammonium hydroxide. This is sprayed directly on the completed chromatogram or electropherogram. A second method is to prepare the 8-hydroxyquinoline as a 1.0 per cent (v:v) solution in a mixture of 80.0 per cent (v:v) reagent alcohol and 20.0 per cent (v:v) distilled water. This is sprayed and the electropherogram or chromatogram is then placed in a container with ammonia vapor and examined under ultraviolet light. Either method will produce characteristic brown-yellow color. With the UV method, a very dark color without fluorescences will be observed.

The second method utilizes a 0.1% aqueous solution of Alizarin S spray followed by exposure to ammonia vapor. Alternately, a saturated solution of the Alizarin in reagent alcohol can be sprayed with saturated boric acid solution following a period of drying of the Alizarin. Either method will produce a violet color with cobalt.

Cobalt Up-take Studies

A foundation for the experiment ideas which follow is to determine the ability of the test organism to assimilate cobalt and to establish its tolerance level for that element. In many cases, a chemical which will be taken up by a microorganism proves toxic at higher concentrations. For validity, all experiment must stay within the tolerance range.

First select the test organism. Next, make up the basic medium in which it will be grown. An easy way to establish the effective concentration range is to add known quantities of cobalt to a series of eight tubes of culture media. The first tube should contain two micrograms of cobalt per mL of medium. Double the amount of cobalt added in each tube of the series. The final tube should contain 256 micrograms per mL. The test will be enhanced by using different forms of cobalt. Cobalt nitrate and cobalt chloride are highly water soluble. Cobalt oxide is moderately water soluble. Various phosphates and carbonates of cobalt are relatively insoluble.

Zajic (1969) reviewed the earlier literature on cobalt assimilation and tolerance by various bacteria. He pointed out that several workers during the 1950's had found that cobalt tolerance could be varied by different media additives. Those include casein hydrolysate, purine, and pyrimidine bases and various amino acids. Exploration of variables such as these would be sufficient for a very good basic research project.

Microbial Assay For Soil Cobalt

Soils must contain three to five ppm of cobalt for forage crops grown on them to have 0.08 to 0.10 ppm. This seems to be the minimum range for good health without feed supplements in raising animals. Similar concentrations are required for vegetables consumed by humans.

Several variations of this experiment could be conducted. Some of the organisms listed in Table 1.2 should be grown from a standardized inoculum on appropriate media containing various quantities of cobalt. Similar cultures should be grown on the same media with the test soil added. Growth rates obtained from the soilless and soil media should be compared. Quantitative analysis of the cobalt content per mg of dry soil mass could then be made. It is important that the control media contain excess quantities of the other mineral requirements of the test microbes. This will insure that cobalt concentration is the only variable.

An interesting variation of this experiment could be made with *Pseudomonas cattleyecolor* which produces a violet pigment containing cobalt. The quantity and intensity of the pigment is related to the amount of cobalt available to the culture. Use of these bacteria would provide a means for visually estimating relative cobalt concentrations if instruments or other more precise analytic procedures are not available. (HINT!: Most science fair judges will require no more than about ten seconds after seeing this project to start asking about the function of the violet pigment in *Ps. cattleyecolor*. The answer to that one is simple-- none is known.) If quantitative analysis procedures are available, the experiment could easily be extended by using water and various organic solvents to extract the pigment from the bacteria. It could then be broken down by burning and the exact quantity of cobalt analyzed. The project could be made even more interesting to mathematically minded judges by calculating the number of atoms of cobalt required to produce various quantities of the violet pigment.

Microbial Concentration Of Cobalt

Most of the world's production of cobalt comes from central Africa where the element is found in association with various copper ores. Other commercial sources include deposits of cobaltite and skutterudite which are found in Sweden, Canada, and to a lesser extent in Cuba and some of the South Pacific islands. Common leaching and mining processes are not economically feasible when applied to the relatively small quantities of cobalt in other soils and mineral deposits throughout the world. With a constantly increasing demand for cobalt and concern about dependence upon foreign sources, a means of concentrating cobalt from borderline levels becomes more attractive. Processes utilizing microbial concentration could possibly be done in a manner similar to that utilized for copper, iron, and other elements. This experiment is designed to utilize bacteria or fungi to remove cobalt from soils or rocks where it is present in measurable, but relatively small quantities.

Once translocation of the cobalt from the soil to the bacteria has occurred, the bacteria could easily be sacrificed and the cobalt extracted.

Following is a basic general procedure which can be easily adapted to any specific combination of cobalt source, microorganism, and media. The best choices of microorganisms for the experiment would include *Torula utilus*, *Pseudomonas cattleyecolor*, *Streptomyces griseus*, *Streptococcus fecalis*, *Bacillus natto*, and *Lactobacilus lactis*. If a chemosynthetic autotroph is desired, *Thiobacilus thiooxidans* would be a good choice. Sutton and Corrick (1961) found that the addition of sulfur to the growth media would increase leaching of cobalt by this organism.

Prepare the appropriate medium for the test organism in the usual manner. Place it in large culture tubes or small Erlenmeyer flasks and add a quantity of the soil or rock containing the cobalt. After the medium has been autoclaved, inoculate it with a standardized quantity of the test organism. Depending upon the normal generation time of the organism chosen, a small quantity of the culture should be removed and the microorganisms separated by membrane filtration. Dry weight measurement of the organisms can easily be made if they are dried on the filter. Analyze for the amount of cobalt present. Let the culture continue to grow. For some species, cobalt will be a limiting factor. When no further increase in organism numbers can be obtained, the final result can be calculated again on a dry weight basis.

Get Out The Calculator

Many of the experiments described in this section are adaptable to--in fact, need--mathematical applications for completeness. The quantities of cobalt required per cell or per unit of cell mass are absolutely essential for valid analysis. The data obtained in the various experiments can be used for constructing a variety of graphs. Correlations of time with production or utilization factors should be made.

Statistical analysis of the data will have to be adapted to the specific experimental results. Since many of the projects involve quantitative measurements of cobalt utilized or tolerated, along with soil counts or measurements of growth based on cell mass, the coefficient of correlation should be considered as a primary validation procedure. If the data obtained is not numerically valid to support a coefficient of correlation, the more general evaluation by concordance techniques might be used.

Scientific Artistry

The theme of cobalt can easily be carried into the project display. Consider the wide variety of different colors produced by the various valence states and compounds of cobalt. Select some which go well together as the color scheme for the project posters. This is a subtle technique which might not be noticed by some judges until called to their attention. It will, however, probably score in their evaluation of the display when they understand its basis.

It is known that cobalt fixation by some bacteria is suppressed by copper or nickel ions. This experiment will indicate whether cobalt might share a common receptor with copper and nickel within the cells of *Ps. cattleyecolor*. If so, it would be expected that the bacterium would not be able to synthesize its pigment.

Procedure

Obtain a culture of *Ps. cattleyecolor* and grow it on a complete nutrient medium such as nutrient agar. Observe for production of the purple pigment.

Next, make up similar agar to which copper and nickel have been added at a series of concentrations ranging from 1.0 to 1×10^{-5} M of the ions. A convenient form for their addition is as the nitrate salts. After this addition to the nutrient agar, be sure to check the pH and adjust if necessary with dilute hydrochloric acid or ammonium hydroxide.

Pour the agar into sterile Petri dishes and inoculate with the culture. Observe for pigment formation. Base the study on color intensity and time for color to develop.

Another area for observation is that of possible temperature differences. It could be that the enzyme system responsible for the pigment production will function in the presence of the copper and nickel at one temperature but not at another. Temperatures of 20°C and 35°C might be appropriate for trial. An interesting addition to the project could be the inclusion of samples of colored compounds formed by the reaction of cobalt with other elements. It might be even further enhanced by showing the effect of the addition of iron and copper to the solutions from which these compounds are formed. One can find references to the chemistry of such reactions in almost any good textbook.

Section VII: Experiment 2

Bacteria, Pot-holes, And Busted Axles

WARNING! Don't let an English teacher see that title on the project display. "Broken," maybe?

This project is guaranteed to get attention and a sympathetic hearing from any judge who just spent the money he was saving for new golf clubs on a front-end alignment for his car instead, because he ran across an unexpected "chug hole" in the street. Few people realize that these are often the result of bacterial attack at the asphalt-soil interface. This experiment is designed to demonstrate damage to asphalt by various bacteria.

We classify it as a "quickie" because it does not require spending much time doing the actual research. However, plan ahead. It involves a long waiting time for bacterial action to occur.

Background Information

The subject of how much bacteria actually contribute to the breakdown of asphalt pavements has been debated by bacteriologists for many years. Zajic (1969) has a good review of the early research in this area. Check the cumulative index to *Biology Digest* for the last few years to find references to recent and current research about this problem.

Procedure: Analysis Of A Hole

The first step is to find a pot-hole which has a likelihood of having been on a lightly traveled back road. The probability of mechanical damage to the pavement by heavy traffic will be much less in such a location.

This might be a good time to consult with a supervisor at the local roads department. He might be able to suggest a location which has experienced consistent repeated problems over a period of time for no apparent traffic reason. Also, when they learn what you are doing, they might be able to assist by actually obtaining a sample. Assure them that there are no plans to make a hole bigger--only a few grams of damaged asphalt is needed. Get the sample from the bottom of the hole. Be sure that it includes some of the soil base with the asphalt. Two or three pieces of material about one cm in dimension should suffice. Place the specimens in a sterile jar or other container to avoid surface contamination.

Isolation Of Bacteria

The next step in the project will be to attempt the isolation and identification of bacteria associated with the asphalt. Table 2.1 shows the most likely bacterial genera one will encounter and a recommended growth/isolation medium for each.

Table 2.1

Genus of Bacteria	Media for Isolation & Initial Growth
Desulfovibrio	API Sulfate Reducer Medium
Thiobacillus	Thiobacillus Medium
Mycobacterium	DuBose Medium
Nocardia	Nocardia broth
Pseudomonas	Nutrient or Trypticase Soy Broth
Crenothrix	API Iron Bacteria Broth
Gallionella	API Iron Bacteria Broth
Sphaerotilus	API Iron Bacteria Broth

Testing

Obtain some samples of normal asphalt paving with at least a two cm depth of the underlying soil. Cut these into cubes of uniform size. They should be at least one cm in dimension. Saturate the soil portion with one of the bacteria cultures isolated from the pot-hole. Repeat with other samples and all of the different bacteria available. Place each of these in a clean, preferably sterile, container and incubate at room temperature. Add to the project by placing duplicate experiments in an incubator at approximately at 30°C.

Eliminate many natural variables by preparing an artificial "pavement." Obtain some soil typical of the original location. Pack it about one-half inch deep in a small container. An old plastic ice cube tray is ideal for this. Saturate the soil with the bacteria cultures. Allow it to dry to a slightly moist condition. Cover the soil with an asphalt patching compound in a cold mix form available from most hardware or builder's supply stores. Put specimens in a location which has free air movement to permit quick evaporation of the asphalt solvents. Be sure to have some uninoculated tubes for controls. Incubate the specimens for a period of three to six months. A year would be preferable. It would be good to have enough replications so that you could check the condition of the asphalt at three month intervals. Incubation should be at an average summer temperature in your area. Provide a means for periodic wetting and drying of the soil to simulate rain conditions.

At the end of the incubation period, separate the asphalt from the underlying soil of the specimen and examine for evidence of deterioration. Interpret the results by comparison of the inoculated specimens with the uninoculated controls held under similar conditions for the same length of time.

Think about the experiment with reference to natural conditions in an area. The more closely you can simulate the actual road characteristics, the better your project will fare in the judging process. An example of such an approach would be to include experimental specimens inoculated with a mixed culture of all the bacteria obtained from the original pot-hole sample. Another enhancement might include incubating at actual outside temperatures to account for day/night and seasonal variations.

This project as described will not require much actual working time. It could easily be expanded into an extended long term project by carrying out many replications of the tests. Also try using different source specimens and different types of asphalt formulations. Another enhancement would be using bacteria cultures of species which are known to be asphalt degraders.

Section VII: Experiment 3

Cellulase Digestion Of Plant Cell Walls

The supporting walls of plant cells are composed of fibers of various celluloses along with pectin and other material. Modern plant research has many uses for the complete cell without its wall. Such cells are called protoplasts. Releasing the protoplasts undamaged by breaking down the cell wall with enzymes can be tricky. This experiment is a study of how different conditions affect the action of the enzyme cellulase in dissolving the wall.

Materials Required

Biology supply companies offer many different kinds and preparations of cellulases. Your school lab probably already has one or two of the more common.

The simplest, fastest version of this experiment will utilize only a single kind of plant. For thoroughness, it should be conducted with both leaf and stem tissues. The project can be easily extended by using two or more unrelated plants. Include woody and herbaceous types for completeness.

Other supplies required include a microscope, dissecting instruments, and slides. Well slides for hanging drop preparations are somewhat better than standard slides.

Procedure

Enzyme preparations vary widely in their concentration and effectiveness. To determine a starting concentration, refer to the supplier's literature which will usually list an average effective dosage.

Most plant enzymes have an optimum temperature range of 20-30°C. Carry out your experiment at room temperature which is usually within this range. Be sure to note the exact room temperature.

Osmotic pressure effects can be devastating to plant cells and tissues removed from the plant. Prepare and carry out your experiments by placing the specimen in the plant cell osmoticum described in the reagents section of this book.

Prepare your enzyme solutions ahead of time in the concentrations you want to use. A few milliliters of each will be sufficient. Keep the solutions refrigerated until ready for use. Do not attempt to sterilize the enzymes by autoclaving as the temperature will denature them.

Peel or dissect your plant material to obtain thin sections ranging in dimension between two and five millimeters. If possible, try to get sections composed predominantly of a single cell type. Consult a biology textbook or other references for photographs and details about cell types. Put a drop of the osmoticum on a microscope slide, or cover slip if you are using a hanging drop preparation. Immediately after removing the test material from the plant, put it into the osmoticum. Add a measured quantity (0.1 mL is a good starting point) of enzyme solution to the suspension on the slide. Complete the preparation by addition of a cover slip or use a well slide.

If preparing more than one specimen at the same time, allow an interval of three to five minutes between addition of the enzyme to each slide. This is necessary to permit observation of each with accurate standardized time during the early stages of the experiment. Observe the slides for disappearance of the cell walls. These observations should be made at fifteen minute intervals during the first hour after the addition of the enzyme. If results are not apparent, observe 30 minutes later. The time interval can then be extended to an hour for the next three to four hours. Continue that interval of observation if results are evident. If no results are apparent after four hours, extend the observation time to every four or six hours during the next 24 hour period.

Keep a detailed record of results from each observation. Make the experiment more complete if time permits by conducting it at different temperatures of incubation and by using different kinds of enzyme preparations. If you attempt even a moderate number of variations, it can get quickly confusing. This is a place where you should certainly plan in advance and write out a detailed protocol for what you plan to do.

If your school has a video microscope setup, a time lapse or short segment video of the action on the cell walls will make a dramatic addition to your presentation. Remember, however, that the judge is working under a strict time limitation for each project, so keep this short.

Trouble Spot

Some plants have a large amount of pectin in their supporting structure. The pectin might interfere with cellulase penetration. If you have trouble obtaining digestion, try mixing pectinase with the cellulase.

Section VII: Experiment 4

Conditions Affecting Pollen Tube Formation

When a grain of pollen lands on the stigma of a flower, the pollen will start to germinate its tube nucleus. This results in the growth of a tube from the pollen grain down through the style to the ovary of the flower. The sperm nucleus of the pollen grain will divide, develop into a sperm, and move down the pollen tube to fertilize the ovum in the ovary. This project is a study of conditions which might affect the formation of the pollen tube.

Figure 4.1 shows the appearance of a typical pollen tube.

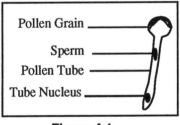

Figure 4.1

If a pollen tube is going to be produced, it will develop very rapidly. Be prepared to watch for the results under the microscope for about one half hour after placing the pollen in the nutrient solution. Estimate the length to which it has developed at various time intervals.

Materials Required

Have all materials prepared and ready to use before obtaining the flower for the pollen. Be ready to use it as soon as possible after removal from the plant. The following will be needed:

Murashige & Skoög Salt Base
Sucrose
Mannitol

Glucose
Microscope and slides
Pointed forceps

Obtain a flower bud at the stage of development shown in Section 6
Experiment 1. Almost any kind will do except some hybrids which
may be infertile. Better results might be obtained by using wild flower
varieties.

Procedure

The choice of microscope and materials should be based on the size of
the flower. With a very small flower, you may need the scanning
objective or even low power of a regular microscope. If this is the
case, make the preparation on an ordinary slide or prepare it as a
hanging drop on a well slide. If the flower is larger, use a watch glass
or a Petri dish with a binocular dissecting scope. A hand held
magnifying glass may be sufficient if you are using an extremely large
flower.

Prepare the nutrient solutions for germination according to the list that
follows. Remember that only a few drops will be needed of each, so
make a small quantity.

1.   **Murashige & Skoög Salt Base**
2.   **Salt base plus 1% sucrose**
3.   **Salt base plus 3% sucrose**
4.   **Salt base plus 3% mannitol**
5.   **Salt base plus 6% mannitol**
6.   **Salt base plus 1% sucrose plus 6% mannitol**
7.   **Salt base plus 3% sucrose plus 6% mannitol**

Use your forceps to remove anthers from the flower. Shake or scrape
off a few pollen grains into each test solution which should be ready on
your slide or in your watch glass.

Work quickly, and as soon as your preparations are ready, begin
observing for development of pollen tubes. They should look
somewhat like the illustration.

Record the time it takes for growth of the pollen tube to start and the time required to reach a specified length. You might find it worthwhile to carry out the experiment once or twice to know what to expect from your particular flower before starting your final measurements. If this is done, be sure to include it in your project report. It will score heavily with the judges.

Extended Experiments

The work described above will be a good project. If you find it interesting and have time, there are several ways that you could develop it into a very large extended study. Here are some suggestions.

Try it with the salt base at one half and one-tenth normal strength.

Substitute sorbitol in place of mannitol.

Substitute glucose for sucrose.

Remove stigmas and styles from some of the flowers, crush them thoroughly, and add a measured quantity of each of your basic nutrient solutions.

Conduct the experiments at different temperatures and light intensities.

Compare the results from different kinds of plants.

Compare the results with pollen obtained at different stages of the flower development ranging from early bud to fully open. Some biologists who have conducted fertilization or androgenesis experiments have reported better results if the flower buds are chilled for a period of 6-24 hours before use. To try this, place buds in a plastic bag which can be sealed to prevent moisture loss and place in the refrigerator for various time periods before use. If the bug of interest for this project has grabbed your attention, it would make a good jumping off point for extension into an androgenesis experiment as described in Section 6. Good luck!

Section VII: Experiment 5

Eat Chocolate Before You Go To Bed?

For generations, parents have cautioned their children not to eat candy before bedtime. The reason given was usually something like a blunt "It will rot your teeth!" Occasionally, the child was further told that sugar and other ingredients in the candy would furnish nutrition for bacteria involved in the process of tooth decay.

If we are concerned only with the sugar content, certainly there is validity in the idea. However many parents, to say nothing of dentists, have experienced shock at some suggestions being made by a few scientists during the last several years. A trip to the library will turn up a number of reports indicating that consuming chocolate just before retiring actually has a beneficial effect. The chocolate is reputed to slow down the growth of some of the bacteria involved in the decay process. A little further research in the library will bring to light the fact that chocolate was used in primitive medicine in areas where the Cacao tree (*Theobroma cacao*) grows. The cacao beans were ground into a powder or extracted by boiling in water and applied to various skin infections. This experiment will show whether there is any validity in either the ancient practice or some of the modern research reports.

Procedure

Remember that most science fairs require prior approval of projects involving bacteria. Be sure your paperwork is in order before you start.

You will need chocolate in various forms such as cocoa, cocoa butter, sweetened and unsweetened bakers chocolate, and various types of chocolate candy. Ideally, you should use some of the mouth bacteria which are implicated in tooth decay. From the practical standpoint, this will probably not be possible because even though we all have these bacteria in our mouths, they are technically considered pathogens, and therefore, would be prohibited by science fair safety rules. You can show the effects of the chocolate by using representative harmless bacteria. Be sure to include both Gram positive and Gram negative bacteria and various morphological types such as bacilli, streptococci, micrococci, and spirilla. Your school probably maintains stock cultures of a number of these. Some good choices might include *Bacillus cereus*, *Streptococcus salivarius*, *Micrococcus lutea*, *Staphylococcus epidermidis*, and *Enterobacter aerogenese*.

Any general purpose medium such as nutrient agar will be suitable for the tests. Prepare the agar according to the manufacturer's directions, sterilize, and pour into sterile Petri dishes. After the agar has solidified, inoculate it by smearing the surface completely and evenly with your bacteria. Use a different plate for each species tested.

A convenient means of smearing the bacteria on the plate surface is illustrated in Figure 5.1. It utilizes a bent glass rod to evenly distribute the inoculum. The glass rods can be prepared and sterilized before hand. Be sure that they are not exposed to the air before you are ready to use them.

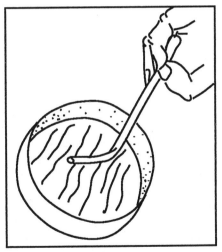

Figure 5.1

Next, place small portions of the chocolate materials on the surface of the plate. Handle the test materials with sterile forceps or spatulas as appropriate. Keep the chocolate test materials at least two cm apart on the plate surface. Figure 5.2 shows how your plate might look at this stage. If you think the chocolate might dissolve or otherwise become unrecognizable after a few hours, use a wax pencil to number each type on the bottom of the plate. That way you can be sure of the identity of the specimen later.

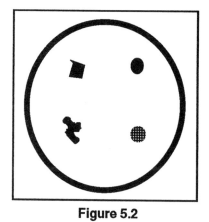

Figure 5.2

If the chocolate does exhibit inhibitory or other harmful activity against the bacteria, they will not grow close to it. Growth in other areas of the plate will be a solid mass. Figure 5.3 shows how a plate might look with one type of chocolate having no effect, another showing a moderate degree of inhibition, and a third showing a large zone of inhibition. Sometimes a testing procedure such as this will reveal a ring of stimulated growth just outside the region of inhibition. This is the result of very low concentrations of the bacteriostatic substance diffusing out to that diameter. The bacteria recognize potential harm in the substance and respond to it by reproducing at a faster rate.

Repeat the set-up with all different varieties of bacteria that you plan to test. After incubation compare results.

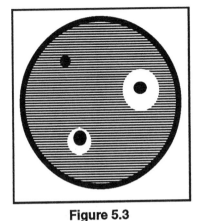

Figure 5.3

If you wish to extend the project, you might want to consider testing the effects of concentration of the materials which had a positive result in the initial screening. To do this, simply take the original chocolate and dilute it to various concentrations. Then apply a given quantity of each dilution to the plate. You can use any dilution method you choose. A convenient means of achieving the dilutions is the serial dilution process which will provide a series of dilutions in which each varies from the previous by a factor of ten. Figure 5.4 shows how to make such a series in screw cap test tubes. Be sure that the diluent is sterilized before you start.

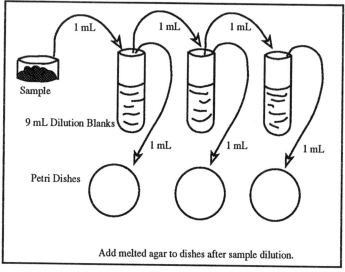

Figure 5.4

Be Prepared

Here are a few topics judges might inquire about. You will profit by being prepared beforehand.

How do you know that the inhibition is the effect of the chocolate and not some other substance? Some candy and other chocolate products contain mold inhibitors.

If a particular candy product did not work, why not? The answer to this one might well be that many of the chocolate candy products on the market are coated with a thin film of paraffin to stabilize them and make handling easier. Such coatings might prevent the chocolate from coming into actual contact with either the agar or the bacteria.

Does a larger zone of inhibition necessarily indicate stronger action? The answer to this one is "No." In some cases it might; in other cases the size of the inhibited zone is simply a reflection of the solubility of the chocolate compound in the aqueous part of the agar. Highly soluble materials will diffuse faster and farther. You might point out that this is something which technologists in a medical laboratory must consider carefully in evaluating the action of different antibiotics on pathogens isolated from patients.

Be sure to take pictures at various stages of your research. Certainly, you will want photographs of the final results. Pictures are particularly important with bacterial projects because you will probably not be allowed to exhibit the actual tests at the science fair. The fact that you will not have lots of physical materials to show will not detract from your score provided your work is otherwise well documented.

Section VII: Experiment 6

Calling Dr. Ape

This experiment will take you to the leading edge of a new science--zoopharmacognosy. The name was coined in 1993 by Jisaka *et al.* who studied the use of medicinal plants by wild chimpanzees. The science of zoopharmacognosy includes the use of many natural remedies and disease preventives by a large number of different animal species.

A number of practical applications involving the use of medicinal plants by lower animals provides a lead to possible new drugs for human use. Extracts of some of the plants used by animals in their natural habitat are already being tested for possible human use in a variety of applications ranging from cancer treatment to worm infestation prevention.

There is little actual laboratory experimentation that can be done with this subject. The importance of it lies in the use of plants for drug purposes by animals in their natural environment. Although the project is observational in nature, since it involves vertebrate animals approval of the appropriate science fair committee should be obtained before starting.

Here are a few suggestions for possible starting points. You will develop many more ideas from extensive reading about animal diseases and behaviors.

One interesting project could be based on the choice of nest materials by birds. Some birds utilize nest linings of leaves and twigs which are known to have insecticidal or insect repellant properties. If your project involves the investigation of birds nests be sure that it is done after the mating and rearing season is concluded and the nest has been deserted. Another project could be based on the use of particular plants by domestic animals such as cattle, dogs, and cats. Could you correlate unusual feeding practices with the apparent physical condition of the animal? For example it is popularly believed that dogs and cats will consume grass and other plant materials only to treat an illness.

Conversations with veterinarians and zoo keepers could furnish additional leads and clues to sources of information.

Section VII: Experiment 7

Color Your Brain

A brief glance at ads in a magazine is sufficient to demonstrate the impact of color. Likewise, watching a few movies will reveal how color is used to help achieve the viewer mood desired by the producer. The effectiveness of color on the mind is beyond doubt; how it works is still the question.

This experiment is designed to investigate how color affects subjective judgement. You can find many studies of color effects in the psychology and neurology journals. This author (Tant, 1995) recently reviewed research relating color to sexual attraction and behavior. The experiment described below is only one of many that could be done to investigate the effects of color on various aspects of human judgement.

Once you have chosen a specific topic and approach, the first step is to be sure the paper work is in order. Human experimentation is involved, so be sure you have the appropriate permission and informed consent forms completed.

Experiments of this type involve statistical analysis. Be sure you have a sufficient number of subjects and tests to be statistically significant. Control or eliminate as many variables as possible, particularly environmental conditions that might affect the results. Think about it and make a list. You might begin with the room in which the tests are conducted. Are they all done at the same time of day? Is the lighting the same? Is the temperature the same? Are all the raters sitting in the same kind of chair? Etc., Etc., Etc.

The basic procedure for this experiment involves judgment of the attractiveness of different persons based upon their photographs. The photos should be similar-- head and shoulder portraits of the same size and quality. The school yearbook files from previous years might be a good source. Preferably, the subjects should not be known by the raters. Black and white pictures, if available, would be better than color.

Set up a rating scale to be used by all participants. For this type of study, many experiments use one like this:

1. Very unattractive
2. Unattractive
3. Neutral - neither attractive nor unattractive
4. Attractive
5. Very attractive

The next step is a trip to the library to find out what is known generally about the effects of different colors on mood. CAUTION! Be thorough in your background preparation. Some judge is going to ask you to explain such things as hue, intensity, and reflectance. The same kind of judge will want to know about male/female differences in color vision and perception. Be prepared, also, to discuss the physics of the visible portion of the electromagnetic spectrum. Judges with a psychology background are likely to ask questions about how the photoreceptors of the eye work and the pathways by which their signals are transmitted to the brain.

You are quite correct if you are getting the impression that this experiment will open up a BI-I-IG can of worms for questions. Add to those above the ones that psychologists will have: "What is attractiveness? Is it the same in all societies? Is it a learned or inherited behavior?" The benefit of all this is that the project will enable you to cover a wide area of knowledge, thereby helping your score.

When you have decided on appropriate colors to use, place each photograph on a colored card as shown in Figure 8. Be sure that a large area of the card is visible around the picture. Mounting the picture in photo mounts will enable you to transfer it to other colored cards as needed.

Identify the pictures by number or letter. Give the raters a form to record their impressions. Limit viewing time to ten to fifteen seconds. When you have finished the viewing phase, total the results and analyze for color background effects.

Several interesting variations could enhance the project. One might be to compare the results of same and opposite sex raters. Another might be a comparison of results obtained when all the raters in the tests room at once are the same sex or mixed. As you get into the project, other possible variations will appear.

The experiment just described is only one of many in which color might affect subjective judgements. Think of others such as the effect of environmental color on music or odor interpretations. With some thought, a,most any of these could be expanded into continuing projects.

Appendix A

Murashige & Skoog Salt Base

Components	mg/liter
NH_4NO_3	1650.000
KNO_3	1900.000
$CaCl_2$ (Anhydrous)	333.000
$MgSO_4$ (Anhydrous)	181.000
KH_2PO_4	170.000
FeNaEDTA	36.700
H_3BO_3	6.200
$MnSO_4 \cdot H_2O$	16.900
$ZnSO_4 \cdot 7H_2O$	8.600
KI	0.830
$Na_2MoO_4 \cdot 2H_2O$	0.250
$CuSO_4 \cdot 5H_2O$	0.025
$CoCl_2 \cdot 6H_2O$	0.025

Murashige Shoot Multiplication Medium A

Components	mg/liter
Salt Base	4303.530
$NaH_2PO_4 \cdot H_2O$	170.000
Adenine Sulfate	80.000
2iP	30.000
IAA	0.300
i-Inositol	100.000
Thiamine HCl	0.400

Murashige Shoot Multiplication Medium B

Components	mg/liter
Salt Base	4303.530
$NaH_2PO_4 \cdot H_2O$	170.000
Adenine Sulfate	80.000
IAA	2.000
i-Inositol	100.000
Kinetin	2.000
Thiamine HCl	0.400

Gamborg, Miller & Ojima Salt Base

Components	mg/liter
$(NH_4)_2SO_4$	134.000
H_3BO_3	3.000
$CaCl_2 \cdot 2H_2O$	150.000
$CoCl_2 \cdot 6H_2O$	0.025
$CuSO_4 \cdot 5H_2O$	0.025
$FeSO_4 \cdot 7H_2O$	27.800
$MgSO_4 \cdot 7H_2O$	250.000
$MnSO_4 \cdot H_2O$	10.000
KI	0.750
KNO_3	2500.000
Na_2EDTA	37.300
$Na_2MoO_4 \cdot 2H_2O$	0.250
$NaH_2PO_4 \cdot H_2O$	150.000
$ZnSO_4 \cdot 7H_2O$	2.000

Heller Salt Base

Components	mg/liter
$AlCl_3$	0.03
H_3BO_3	1.00
$CaCl_2.2H_2O$	75.00
$CuSO_4.5H_2O$	0.03
$FeCl_3.6H_2O$	1.00
$MgSO_4.7H_2O$	250.00
$MnSO_4.4H_2O$	0.10
$NiCl_2.6H_2O$	0.03
KCl	750.00
KI	0.01
$NaNO_3$	600.00
$NaH_2PO_4.H_2O$	125.00
$ZnSO_4.7H_2O$	1.00

pH Control and Measurement

Most biological systems function within a very narrow range of acid-base balance. Therefore, pH of nutrients, wash solutions, etc. is often critical. If you are not already familiar with "pH" and its relation to the hydrogen ion concentration in a solution, you can find good explanations in many standard references such as Hawk's Physiological Chemistry.

Buffers are chemical compounds or mixtures which will ionize in solution to maintain pH within a specific range. Table 1 gives some commonly used biological buffers.

Table 1
Primary Standard Buffer Solutions

Buffer	Composition	g/L	pH @ 20°C
Tartrate	$KNC_4H_4O_6$	Satd. at 25°C	3.557
Citrate	$KH_2C_6H_5O_7$	11.410	3.788
Phthalate	$KHC_8H_4O_4$	10.120	4.002
Phosphate	KH_2PO_4 Na_2HPO_4	3.388 3.533	6.881
Phosphate	KH_2PO_4 Na_2PO_4	1.179 4.302	7.429
Borax	$Na_2B_4O_7 \cdot 10H_2O$	3.800	9.225
Carbonate	$NaHCO_3$ Na_2CO_3	2.092 2.640	10.062

For critical work, the distilled water used as solvent should be purged free of carbon dioxide before making the solution.

pH Indicators

Many chemical compounds change color in different ranges of pH. Table 2 shows the pH range detected by several commonly used materials.

Table 2

pH Indicators

Indictor	pH Range	Color Change
Arid cresol red	0.2-1.8	Red-yellow
Acid thymol blue	1.2-2.8	Red-yellow
Acyl blue	12.0-13.6	Red-blue
Benzo red	4.4-7.6	Red-blue
Benzo yellow	2.4-4.0	Red-yellow
Bromcresol green	3.8-5.4	Yellow-blue
Bromcresol purple	5.2-6.8	Yellow-purple
Bromphenol blue	3.0-4.6	Yellow-blue
Bromthymol blue	6.0-7.6	Yellow-blue
Chlorphenol red	5.2-6.8	Yellow-red
Metacresol purple	7.6-9.2	Yellow-purple
Methyl red	4.4-6.0	Red-yellow
Parazo orange	11.0-12.6	Yellow-orange
Phenol red	6.8-8.4	Yellow-red
Thymol blue	8.0-9.6	Yellow-blue
Tolyl red	10.0-11.6	Red-yellow

Appendix B

Approximate Metric-English Conversions

To Convert	Multiply By	To Find

Length

To Convert	Multiply By	To Find
inches	2.54	centimeters
feet	30.00	centimeters
millimeters	0.04	inches
centimeters	0.40	inches
meters	3.30	feet

Area

To Convert	Multiply By	To Find
square inches	6.50	sq. centimeters
square feet	0.09	square meters
square yards	0.80	square meters
acres	0.40	hectares
sq. centimeters	0.16	square inches
square meters	1.20	square yards
hectares	2.50	acres

Mass

To Convert	Multiply By	To Find
ounces	28.00	grams
pounds	0.45	kilograms
grams	0.04	ounce
kilograms	2.20	pounds

Volume

fluid ounces	30.00	milliliters
pints	0.47	liters
quarts	0.95	liters
gallons (US)	3.80	liters
gallons (imp.)	4.50	liters
cubic feet	0.03	cubic meters
cubic yards	0.76	cubic meters
milliliters	0.03	fluid ounces
liters	2.10	pints
liters	1.06	quarts
liters	0.26	gallons (US)
liters	0.22	gallons (imp.)
cubic meters	35.00	cubic feet
cubic meters	1.30	cubic yards

Note: These are for quick approximation only. Consult appropriate references if accuracy beyond 1 or 2 decimal places is needed.

Appendix C

Sources of Supplies

Central Scientific Co.
3300 CENCO Pkwy.
Franklin Park, IL 60131-1364
(708) 451-0150

Connecticut Valley Biological Supply Company
P.O. 326
Southampton, Mass. 01073
(413) 527-4030

Edmund Scientific Co.
101 East Gloucester Pike
Barrington, NJ 08007-1380
(609) 573-6250

Flinn Scientific
P.O. Box 219
Batavia, IL 60510-9906
(708) 879-6900

Nasco
901 Janesville Avenue
Ft. Atkinson, WI 53538
1-800-558-9595

Nebraska Scientific
3823 Leavenworth Street
Omaha, Nebraska 68105
1-800-228-7117

Sargent-Welch Scientific
7400 North Linder
Skokie, IL 60076
1-800-727-4368

Sargent-Welch Scientific, Canada
77 Enterprise Drive North
London, Ontario N6N 1A5
1-800-265-3496

Science Kit & Boreal Laboratories
777 East Park Drive
Tonawanda, NY 14150-6784
1-800-828-7777

Sigma Chemical
P.O. Box 14508
St. Louis, MO 63178-9916
1-800-325-3010

Synthephytes
P.O. Box 1032
Angleton, TX 77516-1032
(713) 369-2044

Appendix D

Abbreviations

AA- Amino acid.

ABA- Mucopolysaccharides responsible for the ABO blood group system; also, Abcissic acid.

BA- Benzyl adenine.

CMS- Cytoplasmic male sterility.

pCPA- para-chlorophenoxyacetic acid.

2,4-D- 2, 4-dichlorophenoxyacetic acid, a phytohormone used as a weed killer.

DNA- Deoxyribonucleic acid.

cDNA- complementary DNA produced from a RNA template by the action of RNA- dependent DNA polymerase (reverse transcriptase).

pDNA- Plasmid DNA.

rDNA- In general, any DNA regions that code for ribosomal RNA components.

tDNA- A group of seven genes of the Ti-plasmid that integrate into the nuclear DNA of the host plant during tumor induction.

ELISA- Enzyme Linked Immunosorbent Assay

GA- Gibberellic acid.

GMO- Gamborg, Miller, & Ojima media.

IAA- Indole acetic acid.

2iP- 2-(isopentenyl)-adenine.

IRB - Institutional Review Board

K or **KIN**- Kinetin.

MS- Murashige & Skoög.

NAA- Naphthalene acetic acid.

PAGE- Polyacrylamide gel electrophoresis.

PBS- Phosphate buffered saline.

PEG - Polyethylene glycol

PCV- Packed cell volume.

P protein- Phloem specific proteins formed following injury.

RIA- Radioimmunoassay.

SH- Schenk & Hildebrandt medium.

SRC - Scientific Review Committee

TC- Tissue culture.

Ti- Tumor inducing. Usually refers to a plasmid from *Agrobacterium tumefaciens*.

Measurement Abbreviations

The abbreviations in this table are universally recognized and may ordinarily be used without definition.

gram	g
milligram	mg
kilogram	kg
liter	L
milliliter	mL
mole	mol
meter[1]	m
centimeter[2]	cm
millimeter	mm
micrometer[3]	μm
density	D
mass	m
volume	V
temperature	T
second	s
Fahrenheit	°F
Celsius	°C
Kelvin	K

[1] Squared or cubed measurements of length are indicated with a superscript 2 or 3, as mm^2, cm^3, etc. The   previous use of prefix abbreviations sq. and cu. is to be avoided.

[2] Be careful not to interchange cubic centimeter and millimeter. A mL is a measure of volume; a cm^3 is   a measure of space.

[3] The older term, micron, designated by the Greek μ, is now to be avoided.

Appendix E

Chromosome Stain

Best results will be obtained from rapidly dividing cells. The choice of stain, time, and fixative technique will vary with the species and physiological state of the cells. Expect to make numerous trial-and-error determinations with any specific system. The following is a good starting point.

Cell Preparation

Cell cultures can be fixed in Farmer's Fixative without further preparation after removal by filtration or centrifugation from the growth media.

Cells from tissues can be obtained by hydrolysis of the tissue in HCl and/or pectinase. Concentrations will have to be determined experimentally for a particular tissue. You can find many procedures applicable to different plants in recent issues of Plant, Cell, Tissue, and Organ Culture.

Fixation And Hydrolysis

Farmer's Fixative is prepared by mixing glacial acetic acid with reagent alcohol, 1:3 (V:V). **CAUTION!** Use eye and rubber gloves in hood.

Leave cells in fixative 2-12 hours. Remove by centrifugation and rinse in distilled water.
Hydrolyze cells by suspending in 1N HCl for 2-6 hours at room temperature. Exact time is cell-type dependent.

Remove cells by centrifugation and rinse twice with 0.1M sodium acetate. Remove cells by centrifugation.

Staining

Place cells in 1.0 to 2.0 mL aceto-camine stain solution for 6-18 hours. Place a drop on a microscope slide, squash with a coverslip, and seal the coverslip with melted paraffin.

(Note: you can find directions for preparing the stain solution. It involves boiling glacial acetic acid, so we recommend against doing it yourself. Prepared stain is available from lab suppliers and is stable for at least a year.)

If you do not obtain acceptable results with the aceto-carmine method, you might wish to investigate obtaining SRC approval to use the Feulgen or Carbol Fuchsin techniques.

Biuret Test For Protein

Conduct the test against a known standard protein. A commonly used one is bovine serum albumin (BSA) at various dilutions of a beginning concentration of 2.0 mg/mL distilled water.

A convenient working volume is 1.5 mL protein solution to which is added 1.5 mL Biuret Reagent. Mix the Biuret Reagent and protein solution thoroughly and incubate for 15 min at 37°C. Read in a spectrophotometer at 540 nm. Draw a reference curve and plot A_{540nm}. Eight to ten concentrations of BSA should suffice for most applications.

Prepare the Biuret reagent as follows:

$CuSO_4$ * $5H_2O$1.5 g
Sodium potassium tartrate6.0 g
10% NaOH (W:V)300.0 mL

This is a caustic solution.   Handle appropriately!

Ninhydrin Test For Amino Acids

The detection of amino acids is easily accomplished by means of the ninhydrin test. The reagent can be applied as a spray to chromatograms on which amino acids have been separated. or it can be applied directly to solutions or even solid materials. The most common use is as a spray reagent for chromatograms. Following separation of the amino acids the chromatogram is dried sprayed with the reagent and heated at approximately 50 °C for ten minutes. Free amino groups will produce a blue to purple colored spot.

Preparation Of Ninhydrin Reagent

Ninhydrin 0.03 g.

n-butanol 10.0 mL.

Dissolve the ninhydrin completely in the butanol.

Then add 0.3 mL glacial acetic acid.

Mix.

Iodine Test For Starch

The reaction of starch with iodine produces a blue-black color. Simply place a drop of the iodine solution in contact with the substance to be tested.

Iodine Solution

Potassium iodide 10.0 g
Iodine 3.0 g
Distilled water 1.0 L

Store in a dark bottle.

Caution. Dust and vapor are toxic. Handle with gloves in hood when preparing.

Glossary Of Terms

Biology, especially plant biology, has changed. New theories and concepts, new facts, and new processes have resulted in a proliferation of new technology. To anyone who has not yet had the opportunity to follow recent developments in this technology, the new language is a confusing foreign tongue. New abbreviations are being rapidly coined for the new words and processes. Unfortunately, some editors of scientific publications are permitting the use of these abbreviations without proper definition.

All this makes it even more difficult for busy persons to keep abreast of the new developments. If we are unable to communicate our new knowledge, can it have any real and lasting value?

Adventitious- Refers to the development of plant organs from unusual origin such as callus tissues.

Aerobic- utilizes oxygen

Androgenesis- The development of a haploid male plant from a pollen grain.

Anerobic- exists in the absence of oxygen.

Aneuploidy- The condition in which the number of chromosomes differs from the haploid number or multiples of the haploid number.

Browning reaction- The occurrence of a brown color in freshly cut tissues or tissue culture media usually as a result of the production of phenolic compounds.

Callus- An undifferentiated group of cells that forms as a response to plant tissue injury or hormone imbalance.

Cell suspension culture- A tissue culture technique involving the multiplication of cells suspended in a liquid medium.

Cytokinesis- The division of the protoplasm which ordinarily follows mitosis.

Differentiated- The development of specialized cells and tissues having specific functions from undifferentiated meristematic, callus, or proembryonic cells.

Explant- The part of a plant removed from the parent and used to initiate *in-vitro* culture.

Genome- The total number of genes ie., the genetic make-up of the organism. It is usually considered to be the haploid number of chromosomes.

Habituation- The condition in which after a number of subcultures, plant tissues adapt to grow without hormones or other specialized nutrients originally needed.

Halophilic- requiring high salt concentration.

Heterokaryon- A cell which has two or more different nuclei as a result of cell fusion.

Homokaryon- A cell which has two or more identical nuclei as a result of cell fusion.

In-vitro- A term used to describe biological processes occurring under artificial laboratory conditions.

In-vivo- A term describing biological processes occurring in living organisms.

Irradiance- The total radiation, including visible and invisible, falling on a surface.

Meristem- The rapidly dividing cells in the growth tips or other regions of a plant.

Micropropagation- Cloning or other vegetative propagation of plants *in-vitro*.

Mordant- a substance, often metallic, that combines with an organic dye to make it insoluble from the material to which it is applied.

Pathogenic- capable of causing disease.

Plating efficiency- Refers to the percentage of inoculated cells which develop into cell colonies.

Primary culture- The initial growth resulting from an explant of cells, tissues, or organs.

Regeneration- The regrowth or replacement of differentiated tissues or organs which have been removed from a plant.

References

Beerstecher, Ernest Jr., 1954. Petroleum Microbiology. Elsevier Press, Inc., New York.

Bennett, E.O., 1956. A discussion of current research in the bacteriology of soluble oil emulsion. Lubrication Eng., J u n e , 1956.

Dixon, R.A., 1985. Plant Cell Culture--A Practical Approach. IRL Press, Ltd., Oxford, England.

Feigl, Fritz, 1978. Spot tests in Organic Analysis. Elsevier Pub. Co., New York.

Fowke, L.C. and F. Constable, 1985. Plant Protoplasts. CRC Publishing Co., Boca Raton, Florida.

Nitsch, Colette, 1984. Production of Isogenic Lines: Basic Technical Aspects of Androgenesis. In Plant Tissue Culture: Methods and Applications in Agriculture, Trevor A. Thorpe, editor. Academic Press, Inc., New York.

Oser, B.L., 1969. Hawk's Physiological Chemistry, 14th Ed. McGraww-Hill, Inc. New York.

Sulton, J.A. and J.D. Corrick, 1961. Bacteria in mining and metallurgy. U.S. Bur. Mines, Rept. Invest. 5839: 1-16.

Tant, Carl, 1991. Plant Biotech--The New Turn-On. Biotech Publishing, Angleton, TX.

Tant, Carl, 1993. Plant Biotech Lab Manual. Biotech Publishing, Angleton, TX.

Tant, Carl, 1994. Awesome Green: The Explosive New Plant Sciences. Biotech Publishing, Angleton, TX.

Tant, Carl, 1995. Awesome Neurochemicals: The Essence of Sex. Biotech Publishing, Angleton, TX.

Thorpe, Trevor A., (Editor), 1981. Plant Tissue Culture Methods and Applications in Agriculture. Academic Press,   New York.

Williams, David B., 1995. Writing Science Research Papers - An Introductory Step-by-Step Approach to A's. Biotech Publishing, Angleton,TX

Wolzogen Kuhr, C.A.H. von and L.S. van der Vlught, 1934. *Bacterial corrosion in water lines.* Water (Netherlands)   18:147-165.

Yamagata, N. and Y. Murakami, 1958. A cobalt accumulator plant. Nature 181: 1808

Young, R.S., 1960. Cobalt - It's Chemistry, Metallurgy, and Uses. A.C.S. Monograph. Reinhold Publishing Co., New York.

Zajic, James E., 1969. Microbial Biogeochemistry. Academic Press, New York.

At Last!

Writing Science Research Papers: An Introductory Step-by-Step Approach to A's

by David B. Williams

Many excellent style manuals are available for graduate students and professional scientists. This book by David Williams fills the need for a "how-to" on scientific writing for pre- and beginning college students. It started as a guide for his own students. He has expanded it to include research projects reports (such as science fairs), topic reports, and literature reviews. Williams lays a clear path through the jungle of scientific writing and even includes a special chapter for middle school students. Detailed explanations of indexes, references, and other library resources are accompanied by multiple examples of note-taking, citation, and bibliography styles. English teachers will value a quick reference to the conventions of different scientific discipline styles.

David Williams has taught Jr. And Sr. High School science for 15 years since receiving a B.S. In education from Southern Methodist University. He is currently teaching high school science in Manassas, Virginia while completing a masters degree in chemistry at George Mason University. Lack of basic introduction to scientific writing inspired this effort to fill a void.   **ISBN 1-880319-17-9 112 pages Paperback $14.95**

Other Books by Carl Tant from Biotech Publishing

Seeds, etc....

Unique experiments with seeds motivate upper elementary and middle school students to think, plan, and interpret. Includes extensive background information and tips for parents and teachers. Although the safest possible materials are utilized, the basics of good laboratory practice are emphasized. The projects avoid a "Cookbook" approach and grasp the student's interest with fascinating ideas not found in common lab manual procedures. Many are adaptable for use as teacher demonstrations for younger students or provide a means for parents to work with their children. Some experiments are ideal for use by students with disabilities or special educational needs. Suggestions are included for extending appropriate experiments into Science Fair or other major projects. Paperback, 160 pp. $13.95. ISBN 1-880319-01-2.

Projects: Making Hands-On Science Easy

Here is the answer to problems with science project assignments. New requirements for more hands-on lab work have caught many teachers and parents ill-prepared to deal with the techniques involved. Author Carl Tant gives vital answers and simple methods based on his many years of successful project teaching. Even finding project subjects becomes easy with his methods. A major problem for both teachers and parents is removed by guidelines for when--and when not--to help. How-to suggestions for grading and motivating simplify tasks that are often difficult. Copy masters are included for data, grade, and schedule charts.

"Aunt Julia's Bread" is an added bonus as an example of how to turn ordinary things into science projects. And, this one even gives you something good to eat!

Science project stress turns into science project fun with this _must_ for every teacher and parent. Paperback, 80 pages. $12.95. ISBN: 1-880319-06-3.

Plant Biotech Lab Manual

Catch the excitement of the New Plant Biotechnologies! Extensive background information plus twelve experiments that can be done with a minimum of equipment in advanced high school or introductory college labs. Easy to follow directions and illustrations. Suggested report and data forms may be copied. Full size repro masters are available.

ISBN 1-880319-03-9   Size 5.5"x8.5"   Comb bound $19.95

THE AWESOME SCIENCE OF BIOLOGY SERIES

Explains the real excitement of the new biotechnologies with reviews of the scientific literature for non-specialists. Get the facts without the hype!

The goal of the ASB Series is to explore the new knowledge with a scientifically qualified author who can translate the technical jargon of the research literature into terms understandable by non-specialists. Our only sources of information are interviews with scientists themselves and peer-reviewed scientific literature.

The sources of information are properly cited in the text and a reference list with full details of publication is provided for interested readers.

Watch for announcements of food, marine, and other biotechnology titles soon. The first two volumes open doors to awesome new plants and knowledge about human sexuality.

Awesome Green--The Explosive New Plant Sciences

Should you be worried about recombinant DNA and genetic engineered foods? Explore the myths, fears, and facts about plant biotechnology with scientist/teacher/author Carl Tant. Down-to-earth descriptions with flow charts and diagrams by illustrator Tammy Crask remove the mysteries from the processes.

Preview some of what is coming soon: Tomatoes that taste like tomatoes; Eat a banana to get your next vaccination; Disease resistant plant, no need for pesticides; New drugs for cancer; and biodegradable plastics from plants. Thought provoking questions are raised about social responsibility of high-tech industry. Finally, the biotech industry gets a report card. See who gets the A's and the F's. This first volume of the Awesome Science of Biology Series reviews the literature and explains complex processes for the non-specialists. **ISBN 1-880319-05-5 Softcover $17.95**

Awesome Neurochemicals: The Essence of Sex

The chemical basis of human sexuality is always an interesting--and sometimes controversial--topic. (So is almost everything else related to sexual behavior!) Explore the submicroscopic world of minute molecules produced by nerve cells which control sexual function. Author Carl Tant translates the scientific literature into understandable terms. Separate facts from fiction and superstition. Finally, the last chapter addresses frightening environmental effects on the reproductive system.
ISBN 1-880319-13-6 Softcover $15.95

A Note About The Author:

Carl Tant recently took early retirement from public school teaching after 21 years as a biology teacher and science chair at Angleton High School. He received B.S. and M.S. degrees in biology from the University of Houston and worked in public health bacteriology and industrial research for several years before starting his teaching career. Now devoting most of his time to writing and administration of a plant biotechnology lab, he also continues to teach as a part-time instructor in biology for Alvin Community College.

And When 10,000 Words Is Not Enough...

Illustrations by Tammy Crask simplify involved processes and procedures. Tammy, a 1994 graduate of The Art Institute of Houston, majored in graphic design. In addition to her technical illustrations, readers will especially enjoy the way her cartoons drive a point firmly home with a smile.

Appendices

Index

Colophon

Text entry, charts, and layout were done with **WordPerfect 5.1**.

Original text entry and data preparation were done by Renee' Walker.

Layout, design, type selection, and preparation of camera-ready copy were done by Tammy Crask. Final print-out was with a **Texas Instruments Microlaser** printer operating in **Postscript** mode.

The front cover illustration was prepared by Tammy Crask as a model of a science fair display. The model was photographed on **Kodak**™ **Technical Pan Film** and printed in high contrast for reproduction.

Tradenames are designated by bold face type in the text. Products so designated are well known and usually held in high public regard; consequently, their use is for the benefit of their owner and does not imply a generic group of products.

Printed in the USA by
BookMasters, Inc.
Ashland, Ohio